Practical Dressmaking

By Morang'a Erick Moseti

Practical Dressmaking

Designed and published by:

Kipchumba Foundation

P.O. Box 25380 – 00100 Nairobi, Kenya

www.kipchumbafound.org

©2019

Morang'a Erick Moseti

No part of this work can be copied, translated or disseminated without the express permission of the author or the publisher. The ideas presented in this book are solely by the author and do not reflect the position of the publisher.

Dedication

I dedicate this book to my wife, daughter, son and my friend Paul Kipchumba whose support has made this book successful.

Acknowledgment

This book is an illustrated text meant to prepare the trainee to become a skilled dressmaker, enable interpretation of various garment designs, and application of concepts learned. Every chapter's content entails an introduction, practice, supplementary content, and some revision questions for testing the trainee's grasp of the content. However, the book does not cover technologies in dressmaking for individual and corporate/ industrial application. There is some concerted effort to produce a book on applied dressmaking that will cover essential but omitted content.

In this respect, I would like to acknowledge my family wife Agnes, children Ignatius and Clare, dad Philip, mum Lilian and my siblings Richard, Duncan and Beatrice, for their support, motivation, enthusiasm, and prayers that played a great role towards accomplishing this work.

In a special way I would like to thank my great friend and founder of Kipchumba foundation Mr. Paul Kipchumba for guidance, encouragement and untiring commitment to this work. My appreciation also goes to my friends and classmates Job Mwangi, Job Samoka for their assistance and advice.

Morang'a Erick Moseti

Kilimambogo, Kiambu County, Kenya

July 2019

Content

Dedication..3

Acknowledgment..4

Abbreviations and Acronyms...11

Chapter 1: Introduction to Dressmaking..................................12

    1.1 Description and Importance of Dressmaking..................12

    1.2 Workplace Organization, Safety and Precautions..........13

    1.3 Tools and Equipment..16

        1.3.1 General Equipment..16

        1.3.2 Sewing Equipment..16

        1.3.3 Pressing Equipment...23

        1.3.4 Sewing Machine..25

    1.4 Textile fibers...35

        1.4.1 Natural fibres...37

        1.4.2 Man-made fibres...45

    1.5 Revision Questions...53

Supplementary Content..57

    1. Personal Appearance..57

    2. Fabrics for everyday garments...64

Chapter 2: Garment Construction..66

    2.1 Taking measurements...66

    2.2 Garment design...70

    2.3 Pattern alteration...72

        2.3.1 To Reduce the Measurements of a Pattern............72

        2.3.2 To Enlarge the Measurements of a Pattern...........74

        2.3.3 Use of a Paper Pattern...76

        2.3.4 Preparation of the fabric..76

    2.3.5 General layout of a pattern………………………………....76
    2.3.6 Marking out the Pattern…………………………………...79
    2.3.7 Preparation for Fitting……………………………………...80
  2.4 Garment construction procedure……………………………82
    2.4.1 Order of Work………………………………………………..82
    2.4.2 Stitches………………………………………………………..82
  2.5 Revision Questions……………………………………………93

Supplementary content…………………………………………....96

  1. Measurement conversions………………………………….....96
  2. Table of measures……………………………………………...97

Chapter 3: Seams……………………………………………….100

  3.1 Introduction…………………………………………………..100
  3.2 Types of seams………………………………………………..101
    3.2.1 Plain Seam: inconspicuous………………………………..101
    3.2.2 French Seam: inconspicuous……………………………...105
    3.2.3 Overlaid Seam: conspicuous……………………………...107
    3.2.4 Machine and Fell: conspicuous…………………………..108
  3.3 Revision Questions……………………………………………110

Chapter 4: Disposal of Fullness……………………………….....111

  4.1 Darts…………………………………………………………...111
    4.1.1 Single Pointed Dart………………………………………..111
    4.1.2 Double Pointed Dart………………………………………112
    4.1.3 Dart Tack…………………………………………………....114
  4.2 Pleats………………………………………………………….115
    4.2.1 Knife Pleats…………………………………………………116
    4.2.2 Box Pleats…………………………………………………..116
    4.2.4 Kick Pleats………………………………………………….118
  4.3 Tucks…………………………………………………………..120
    4.3.1 Markers……………………………………………….….....120

    4.3.2 Wide Tucks...120

    4.3.3 Pin Tucks...121

    4.3.4 Decorative Tucks..122

4.4 Gathering..124

    4.4.1 Gathering by Hand.....................................124

    4.4.2 Gathering by Machine...............................124

4.5 Smocking..124

    4.5.1 Preparation for Smocking..........................125

    4.5.2 Smocking Stitches......................................127

4.6 Revision Questions..128

Chapter 5: Crossway Strips, and Facings and Interfacings.......129

5.1 Crossway Strip..129

5.2 Cutting and joining crossway strips........................130

    5.2.1 Cutting Crossway Strips.............................130

    5.2.2 Joining a Crossway Strip............................130

    5.2.3 Crossway Strips used for Binding..............131

    5.2.4 Crossway Strips Used for Facing...............134

5.3 Interfacings...135

5.4 Facing..137

    5.4.1 Facing on Armhole......................................137

    5.4.2 Facing a neck edge......................................139

    5.4.3 Facing a square neck..................................140

    5.4.4 Decorative neck facing...............................141

5.5 Linings..141

5.6 Revision Questions..142

Chapter 6: Openings..143

6.1 Introduction...143

6.2 Faced opening..143

6.3 Bound opening..145

6.4 Continuous strip opening……………………………………..147

6.5 Revision Questions……………………………………………..149

Chapter 7: Fastenings………………………………………………….150

   7.1 Introduction……………………………………………………150

   7.2 Zip fasteners…………………………………………………...152

      7.2.1 Visual Method: for inserting a zip into a panel without a seam……………………………………………………………..153

      7.2.3 Semi-concealed Method: for inserting a zip into a plain seam…………………………………………………………....155

      7.2.3 Concealed Method: for inserting a zip into a plain seam.157

      7.2.4 Invisible Method…………………………………………158

   7.3 Fastening with Buttons……………………………………….160

      7.3.1 Worked buttonholes……………………………………...163

      7.3.2 Bound buttonholes……………………………………….164

      7.3.3 Worked loops…………………………………………….166

      7.3.4 Rouleau loops……………………………………………168

      7.3.5 Buttons…………………………………………………...170

   7.4 Press Studs…………………………………………………....172

   7.5 Hooks and eyes……………………………………………….172

   7.7 Revision Questions…………………………………………...173

Chapter 8: Pockets and Collars………………………………………..174

   8.1 Pockets………………………………………………………..174

      8.1.1 Patch pockets…………………………………………….174

      8.1.2 Welt Pockets……………………………………………..176

      8.1.3 Jetted pocket……………………………………………..177

      8.1.4 Side pocket……………………………………………….177

   8.2 Collars…………………………………………………………177

      8.2.1 Making of a straight collar……………………………...178

      8.2.2 Peter Pan and shaped collars……………………………186

8.3 Revision Questions..**189**

Chapter 9: Sleeves and Cuffs..**191**

9.1 Sleeves..**191**

9.1.1 Set-in Sleeves...**191**

9.1.2 Raglan Sleeve..**195**

9.2 Cuffs...**196**

9.2.1 Cuff with an overlap...**197**

9.2.2 Open or shaped cuff..**198**

9.3 Revision Questions...**202**

Chapter 10: Waistbands and Belts, and Hems....................**203**

10.1 Waistbands..**203**

10.1.1 Making Waistbands..**204**

10.1.2 Setting on waistband...**205**

10.2 Petersham bands..**207**

10.3 Belts and belt carriers..**208**

10.4 Hems..**210**

10.4.1 Straight Hemlines: finishes.................................**211**

10.4.2 Circular Hemlines: finishes.................................**213**

10.4.3 Flared Hemline: finishes.....................................**214**

10.5 Revision Questions..**214**

Chapter 11: Decorative Finishes..**215**

11.1 Shell edging..**215**

11.2 Pin stitch...**216**

11.3 Faced scallops..**218**

11.4 Faggoting..**219**

11.4.1 Rouleau..**219**

11.4.2 Frills..**221**

11.4.3 Lace..**222**

 11.4.4 Braid...**224**

 11.5 Revision Questions..225

Supplementary content..226

 1. Care and maintenance of clothes............................**226**

 2. Costing (material requirements and labour charges)......**228**

 3. Computer-generated design and pattern making..........**230**

Chapter 12: Garment Making Practical..............................231

 12.1 Skirt...231

  12.1.1 Straight skirt...**231**

  12.1.2 Front skirt..**233**

  12.1.3 Back skirt...**233**

  12.1.4 8 piece panel skirt..**236**

 12.2 Blouse...239

 12.3 Gents Shirt..246

 12.4 Trouser..255

 12.5 Revision Questions..265

Trade Terms...266

Additional Revision Questions......................................268

References..273

Answers to Revision Questions......................................274

About the Author...282

Abbreviations and Acronyms

CAD	Computer-Aided Design
C.B./ CB	Centre Back
C.F./ CF	Centre Front
R.S./ RS	Right Side
W.S./ WS	Wrong Side

Chapter 1: Introduction to Dressmaking

1.1 Description and Importance of Dressmaking

Dressmaking involves planning, drafting, cutting and finishing a piece of garment. With the knowledge of sewing and dressmaking, one can turn to it and earn a good income, at almost a moment's notice.

There are three ways in which you can turn your knowledge into money: draft patterns for others, go out by the day to sew, or open an establishment of your own. If you do the latter you can easily combine the second with it.

To open an establishment of your own, you may set aside certain rooms in your own home for your work, or you may open rooms in some convenience in the downtown district. In one way the latter is best. You can get completely away from your work when working hours are over. The expense, however, is greater.

For your work you should have three rooms: a waiting-room, a fitting-room, and a stitching room. The first should be small but comfortable, and in good taste. The second may be small, but should have a good light and, if possible, a mirror coming nearly to the floor. The stitching room should be lit, warm and large. In this room should be kept all working tools and equipment. You should have in this room a gas cooker or a stove for heating flat irons. A good plan is to tack a large piece of muslin on the wall in this room to pin patterns on. This will be found better than to fold them. Keep all patterns of regular clients. If possible have a closet off of this room to hang your finished and partly finished work in. Have two or three wooden boxes (cigar boxes) to put all the little utensils in, such as pencils, tracing-wheels, tracing-chalks, etc. Always keep your chart where it will not get broken. Never show the garments you make to anyone but the person for whom they are made. It is not professional. You will lose your customers if you do so. When you are ready to open your establishment, reach all your friends, upon which you have neatly written the word "dressmaking".

Keep your sewing-room as orderly as possible. If you need help, train a person to each part of the business. You must keep a book for your accounts, and put in it everything you furnish for your clients, and the cost of each article. Do not purchase any expensive material for a customer without having first had them make a deposit with you of at least 50% the cost of the article. Send a bill with every piece of work you send out. If this is not paid in one month send another bill. Keep sending them each month until paid. In figuring the cost of a garment be sure to get in every item, and be sure not to forget to add a profit if you are to furnish the material. Go to your dry goods merchant the first thing and get a dressmaker's discount.

You will need a large smooth table, large enough to cut a skirt on, a good sewing-machine, and an ironing-board and ironing-cushion.

1.2 Workplace Organization, Safety and Precautions
Potential Accidents and Their Causes

1. Electric Shock
Electric shock will occur when one comes into contact with exposed to naked electric wires. All live electric wires in the workshop should be insulated.

2. Burns
These might occur as a result of heat producing appliances left unattended in the workshop. Electric iron box should be switched off and disconnected when not in use to avoid such accidents. Charcoal iron box should not be left on the pressing table, instead it should be placed away from items that can catch fire.

3. Falls
One may fall due to slippery floor in the workshop. This may be caused by spilled water, grease or oil. The floor of the workshop should always be dry and free from grease and oil.

4. Obstruction

Accidents in the dressmaking workshop may be caused by blocking walk ways. A workshop should always be arranged in order to facilitate easy movement.

5. Improper Handling of Tools and Equipment

This may cause accidents. All tools must be handled appropriately and they should only be used to perform tasks that they are meant for. All tools and equipment should be kept clean and well organized when not in use.

Causes of Fire in a Dressmaking Workshop

i. Electric Fault

Sparking from electric contacts and inflammable environment may cause fire in a workshop.

ii. Inflammable Substances

Inflammable substances in the workshop may cause fire in case they are not properly stored by not being well secured. They may cause fire if stored close to heat producing appliances in the workshop.

iii. Smoking in the Workshop

Smoking is highly discouraged in dressmaking workshop. If the cigarette is not properly handled and disposed it may cause fire that may destroy the workshop.

iv. Careless Handling of Pressing Equipment

Charcoal iron box is heated by burning charcoal. Thus if the iron box is not properly handled the fire in it might spread to other items in the workshop. On the other hand, both electric and charcoal irons may cause fire if just left on the pressing board while still on. Therefore, great care must be observed.

Ways of Preventing Fire in the Workshop

1. All exposed naked electric wires should be insulated and covered to avoid contact. Electric faults should also be

repaired. Faulty electric equipment should not be used until repaired.
2. All inflammable substances in the workshop should be well secured to prevent them from catching fire. They should also be stored away from sources of heat and heat producing appliances.
3. Smoking in the workshop should be avoided at all cost. Cigarette remnants should be cautiously disposed to prevent them from causing fire.
4. Pressing equipment should be handled with great care since they are among the sources of heat in the workshop and should be switched off when not in use.
5. Avail fire extinguishers in the workshop in case there is fire outbreak. One will be able to use extinguisher to put off and prevent it from spreading.

Types of Fire Extinguishers

1. Foam fire extinguisher - for oil fires
2. Carbon dioxide fire extinguisher – for electric based fires
3. Dry power – for Liquefied Petroleum Gas based fire
4. Sand – for general fires
5. Water – for paper and word fires
6. Fire blankets

Safety Precautions in a Dressmaking Workshop

i   Always use appropriate protective clothing while working in a workshop. This would help to prevent injury in case of an accident. These are
    a) dust coat
    b) apron
    c) closed shoes
    d) nose masks

ii Observe workshop rules and safety procedures regarding tools and equipment in the workshop.

1.3 Tools and Equipment

To obtain good results it is necessary to use the correct equipment and to ensure that all tools are kept in good working order. Do not misuse tools; for example, fine scissors should not be used for cutting, paper or thick fabric.

1.3.1 General Equipment

Cutting table: the cutting table should be at least 54 inches wide, this will allow one to effectively lay patterns and enough width for drafting. The table should also be long enough to accommodate more than one garment and all the tools in use.

Mirror: full length mirror will be of good investment to be used during fitting and making crucial decisions and adjustments.

Box or container: used for storage of small items, trimmings and garments that are still under construction; those that cannot be hanged.

Hanging space: used for hanging garment that is partly finished ready for hemming

Tissue paper: support while machine stitching on very fine fabrics

1.3.2 Sewing Equipment

Scissors: this is one of the very important tools that would ensure that fabrics are cut to required shape and size.

Cutting out scissors: a long pair of scissors measuring up to 23 cm. It has heavy blade and one larger handle for the fingers best for cutting more layers of fabric.

Medium pair: used for trimming seams and hanging threads.

Proper selection, use and care of the instruments are of vital importance on the craft of tailoring and cutting. Some of the instruments in common use are shown in the diagram. Other instruments for special purpose are used in the various branches of the trade.

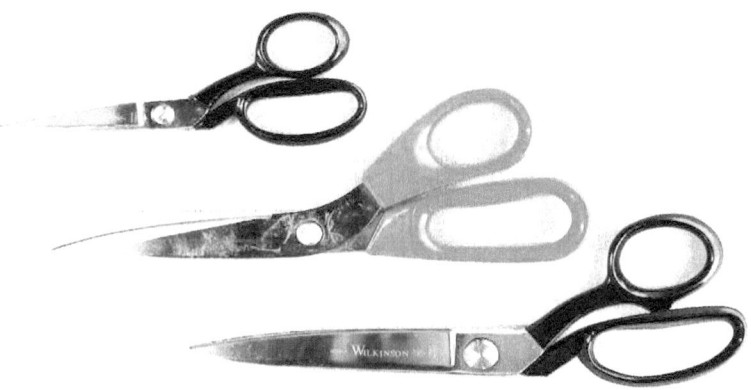

Meter ruler: used for drawing straight lines and for measuring distance. One edge is marked in inches while the other edge may be marked in centimeters.

Tape measure: used for taking body measurements due to its flexibility. Choose a strong, clearly marked measure with metal ends.

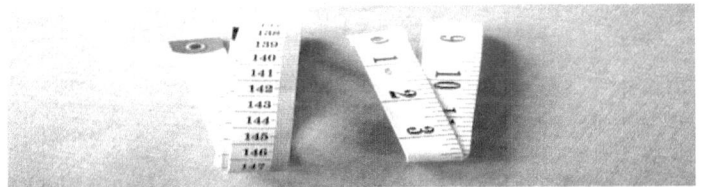

Scale triangle: useful for drawing small diagrams into scale to replicate the actual diagram. Six different scales are printed.

Tailor's chalk: available in different colours for marking lines on fabrics.

Tailor's square: used for drawing right angle lines.

French curve: used for drawing smooth curves.

Pencil: should be well sharpened for drawing diagrams on brown paper and note books.

Rubber: used for erasing errors while drawing or sketching on note book.

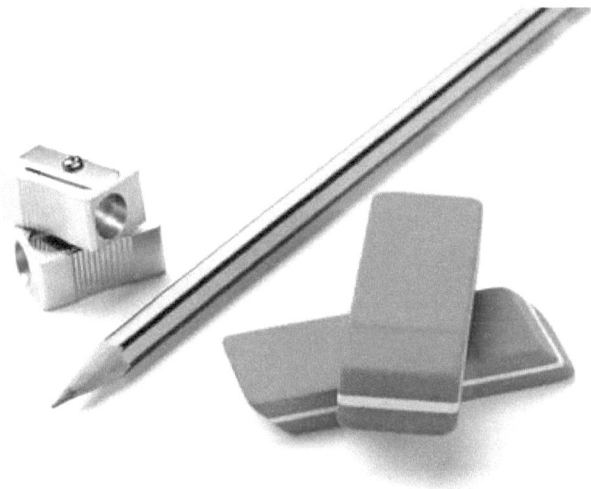

Pinking scissors: they have serrated blades which cut the cloth in a zigzag line. Edges of the cloth cut in this way do not readily get frayed thus eliminating for neatening.

Thimble: is put on the middle finger and is used for pushing the needle when sewing by hand. A steel thimble is the best as this metal is very strong: check that the surface is smooth as rough areas harm fine fabrics. Solid silver thimbles are too soft for constant use. Plastic thimbles crack easily.

Bobbin: used for winding the lower thread of the machine. The Bobbin is placed inside the bobbin-case and then the bobbin case is inserted onto the stud in the centre of the shuttle.

Pins: used for fixing pattern on the fabric for cutting. Pins are also used for holding together two or more layers of fabric ahead of sewing to keep them from slipping away. At the trying on pins are used to tuck up looseness.

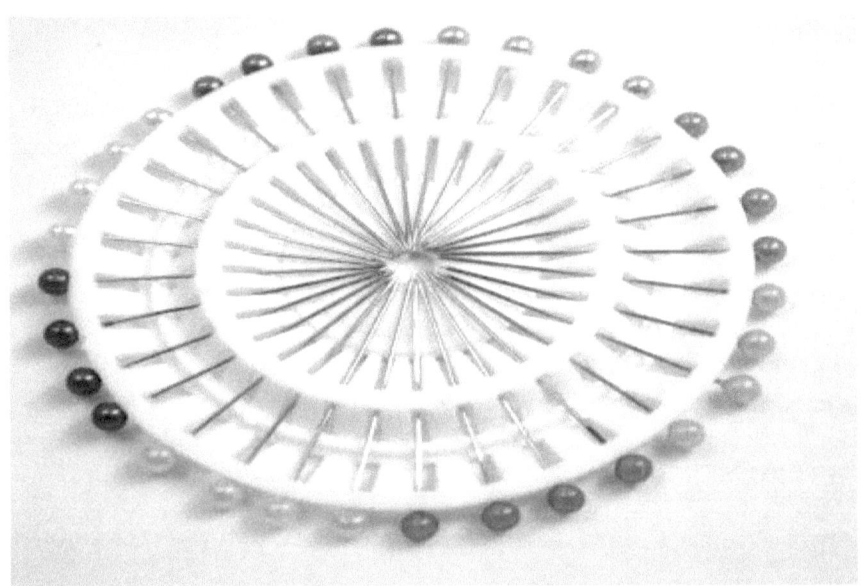

Tracing wheel: used for transferring marks from the paper pattern to another paper placed under it.

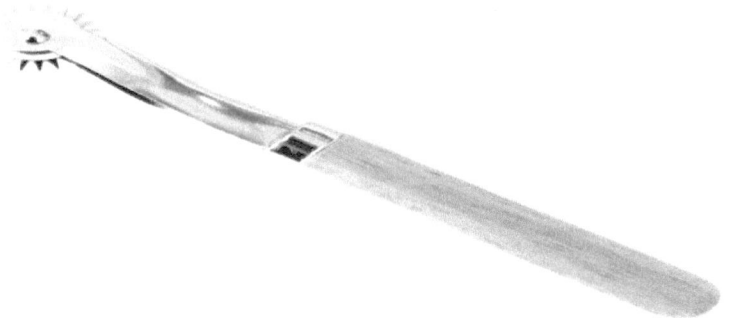

Carbon paper: used for transferring pattern markings and designs.

Seam ripper: used to remove wrongly made and unwanted stitches

Needles

There are many varieties of hand and machine needles for one to choose from. The choice depends on fabric thickness and the type of stitch to be made. Examples of machine needles are no. 16 machine used to stitch thick drill and a no. 14 that may suitable for thin poplin.

Hand needles

Types and sizes vary according to the work to be done and the weight of the fabric used.

Betweens: Short needles with round eyes used for general sewing.

Sharps: Average length needle with a round eye used for general sewing.

Crewel: A needle with a long oval eye for taking more than one strand of thread, similar in length as sharps; used for embroidery but still suitable for most purposes since they are easy to work with.

Darners: Very long with long eyes that make it easy to pick more than one stitch at a time.

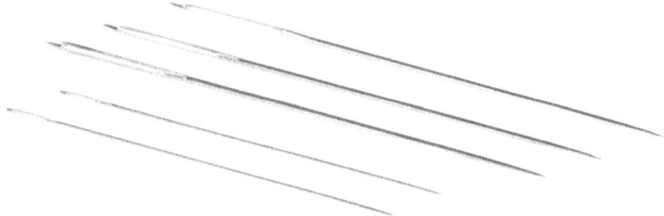

Bodkin: Long blunt-ended needles with a large eye, used for threading elastic through a casing or turning a Rouleau. It is used for pulling out the points of the collar when turning the seam. It is also used for tracing a design with carbon paper.

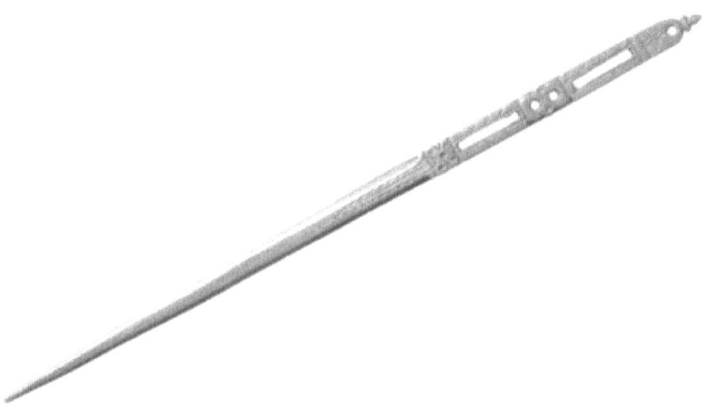

1.3.3 Pressing Equipment

Iron box: used for removing creases on fabrics, fixing iron on interfacing, pressing seams and creating permanent pleats on fabric. It is used for pressing the seams. Ironing gives a smart finish to the garments. Several types of iron box are available and all perform the same function. Electric thermostatically controlled iron box to ensure safe heat for different fabrics. A steam iron has been found to work better.

Ironing board: a flat padded service used for safe pressing or ironing clothes.

Sleeve board: used for pressing sleeves.

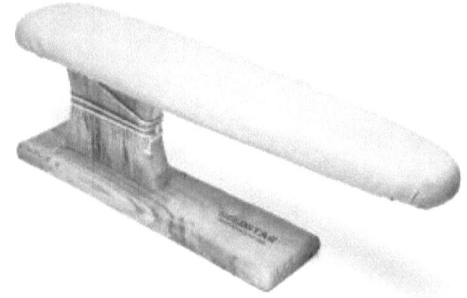

Roller: used for pressing seams.

Pad: Made from a soft blanket encased with cotton fabric; used for pressing sleeve heads and parts that cannot be laid on to the board.

Damping cloths: Made of soft, absorbent cotton fabrics.

Pin board: For pressing velvets and corduroys.

Brief Guide to the Pressing of Fabrics

Cotton: will withstand hot iron with a damp cloth or steam.

Linen: as these fabrics do not shrink, fullness cannot be eased away with pressing.

Silk: moderate iron on the wrong side. Do not use damp cloths as these, and sometimes the use of steam irons, may leave water marks.

Wool: moderately hot iron and damp cloth are essential; fullness can be eased away by shrinkage. Always press wool fabrics under cloth as wool scorches easily.

Rayon and synthetic fabrics: generally a cool iron, without steam.

General Rules

During pressing, a heated iron is gently and firmly pressed down, lifted and pressed down again on to the next section of the fabric. Pressing is necessary to achieve a good finish.

Ironing is where the iron box is moved back and forth on a fabric while slightly pressed down to remove creases.

Pressing darts: press point of the dart smoothly without creasing and press slashed dart open.

Seams: when pressing seam turnings open. Press turnings over a roller so that the impression of the turnings is not created to the right side.

French seams: pressed towards the back of the garment.

Waist seams: press waist seam turnings d upwards.

Overlaid seams: pressed in the direction of overlay.

Sleeve seams: use sleeve board.

Armhole seams: using pressing pad press armhole turnings towards sleeves.

Gathers: push point of iron lightly up into gathers.

1.3.4 Sewing Machine

This is the most costly piece of equipment that one has to buy. It's advisable that one should consult a consumer group or magazine to find the best machine available in the price range. From the many types and brands of sewing machines available one should settle for a machine that can easily be serviced and spare parts readily available.

An Interesting History

In England, when the machine was first introduced (about A.D. 1850) in a big factory, all the workers went on a general strike. In the days before the invention of the machine, all sewing was done by hand alone which required long time and hard work and in effect it had become too costly to make a garment and so the common people used fewer clothes.

Machine's Effect on the Trade

The machine has made a great revolution in the clothing trade. It has decreased the cost of production and increased the supply and so all the people (including the poorest) are better clad today.

In the case of workers, it has reduced the hard labour and the hours of working. Due to the wonderful increase in the demand for clothes, far more people are employed in the trade today.

Various types and sizes

The machines range from a small toy to the giant weighing three and a half tons for belt stitching. They also vary considerably in types and perform almost every kind of sewing, such as darning, embroidery, felling, padding, etc. Button-holes for shirts as well as coats and overcoats are made rapidly. Buttons are also fastened at the speed of ten a minute.

How the stitch forms

Stitch will form when the two machine threads (bobbin and needle thread) form a loop. When the needle passes down the cloth and the sharp point of the shuttle catches the thread to by its semi-circular movement. Thread take up lever pulls the needle thread and stitch formed.

Why the thread breaks

The needle thread will break if the wheel is turns in the wrong direction. Thread would also break when the needle thread tension is too high; loosen the thumb screw to loosen the tension. Test needle thread tension by pulling it through the pressure foot.

Why the needle breaks

The needle will break

(a) If it is not properly inserted with the flat-side inwards.

(b) Needle fixed much lower thus hitting the shuttle.

(c) Needle is not tightened thus it may change its position after some sewing is done and may eventually break.

(d) If thread pulled in a wrong direction after sewing (be pull thread away from the sewer)

(e) Large fabric thinness making it hard for the needle to go through.

(f) If slight bend hitting the side of the needle plate or foot.

Cleaning and oiling

Put a few drops of kerosene and then run the machine speedily to force the dirt out. With a rag wipe the dirt and excess kerosene. Then drop machine-oil at every place of friction.

Parts of a Sewing Machine and Their Functions

The essential structure of sewing machine is similar for machines operated by hand, treadle, or electric sewing machine. The most common parts of a sewing machine are outlined below.

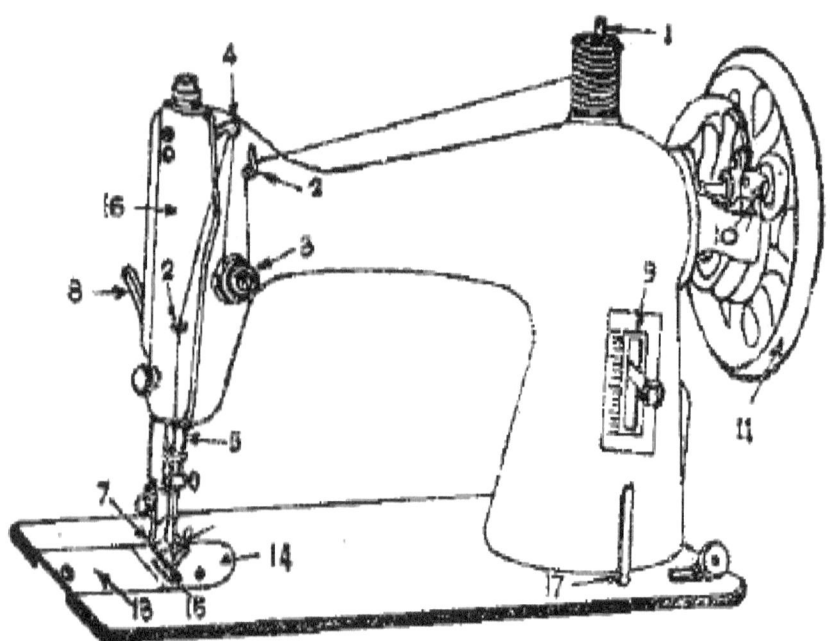

1. ***Spool pin:*** It is situated on top of the arm to contain the needle thread.
2. ***Thread guide:*** It leads and carries the thread in place from the spool to the needle.
3. ***Tension disc:*** it comprises of two concave discs set in contact with each other with the convex sides facing each other. The thread draws between the two. Tension of the needle thread is adjusted by a spring and nut which enhances or reduces pressure
4. ***Take up lever:*** It is a lever attached on to the arm it has up and down motion when machine is running. This makes it provides thread to the needle during its upward motion and tightens the stitch formed during its down ward motion.
5. ***Needle bar:*** This is a steel rod that accommodates the needle at one end with the aid of a needle clamp intended to give motion to the needle.
6. ***Bobbin case:*** This hold the bobbin on which the under thread is wound.

7. **Presser foot:** It is attached to the presser bar to hold the fabric securely in place when brought down.
8. **Presser foot lifter:** A lever connected to the presser bar for lifting and lowering the presser foot.
9. **Stitch regulator:** This regulates the stitch length.
10. **Bobbin winder:** A simple mechanical aspect used to wind thread on to the bobbin.
11. **Hand wheel:** it works the technical aspect running a sewing machine when made to rotate.
12. **Clutch or Thumb Screw:** This is a screw at the center of the hand wheel that engages the stitching mechanism when tightened and loosened to disengage the stitching mechanism.
13. **Slide Plate:** A rectangular sheet of metal, which aids the removal of the bobbin case without raising the machine.
14. **Needle Plate or Throat Plate:** A semi-circular plate with a hollow to allow the needle to go through it while preventing fabric from going through.
15. **Feed dog:** This comprises of a set of teeth fixed below the needle plate. It helps to move the fabric forward when sewing.
16. **Face plate:** A cover which when removed allows access to the oiling areas of the needle bar presser bar and take-up lever.
17. **Spool pin for bobbin winding:** holds thread in place at the time of winding bobbin.

Types of sewing machines

Sewing machines can be divided into four main categories depending on their functionality:

1. **Straight stitch: Hand/ treadle**

These kinds of machines are manufactured in small numbers since they are efficient for commercial use. Stitching is limited to forward and reverse straight stitching of various lengths. It is important to know the make and model number of the sewing machine, as the foot which fits one model will not necessarily fit another.

2. Zig-zag: Electric

These are single needle machines which in addition to stitching straight, the needle moves from left to right to produce a zig-zag of varying length and width as selected. Works overcasting buttonholes, and by closing the stitch length, satin stitch for applique work.

3. Semi-Automatic: Electric

The machine makes zig-zag stitches with four or five stitches are built into the machine.

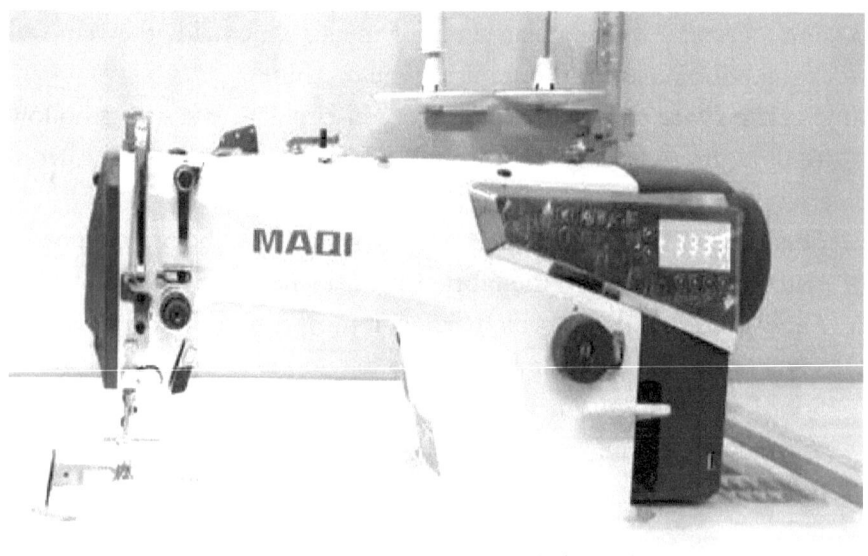

a. Three Step Zig-Zag or Serpentine

Used for overcasting in sheer fabrics to stop fabric from rolling over, woolen fabrics to prevent fraying and sewing on elastic.

b. Blind Stitch

Usually four straight stitches and one zig-zag. Used for either invisible hems or for shell edging on fine fabrics.

c. Elastic Blind Stitch

2/3 small zig-zags and one larger one. Used for invisible hems on knitted or stretch fabrics or for decoration.

d. Multistretch Stitch

Used for seaming together very elastic fabrics

4. Automatics and Superautomatics: Electric, Electronic

These machines are computerized or programmed sewing machines. They can sew the basic stitches with the addition of embroidery patterns. The patterns are produced by changing the programme. Superautomatics, in addition to having a left to right movement of the needle, have movement of the fabric forwards and backwards by the feed teeth.

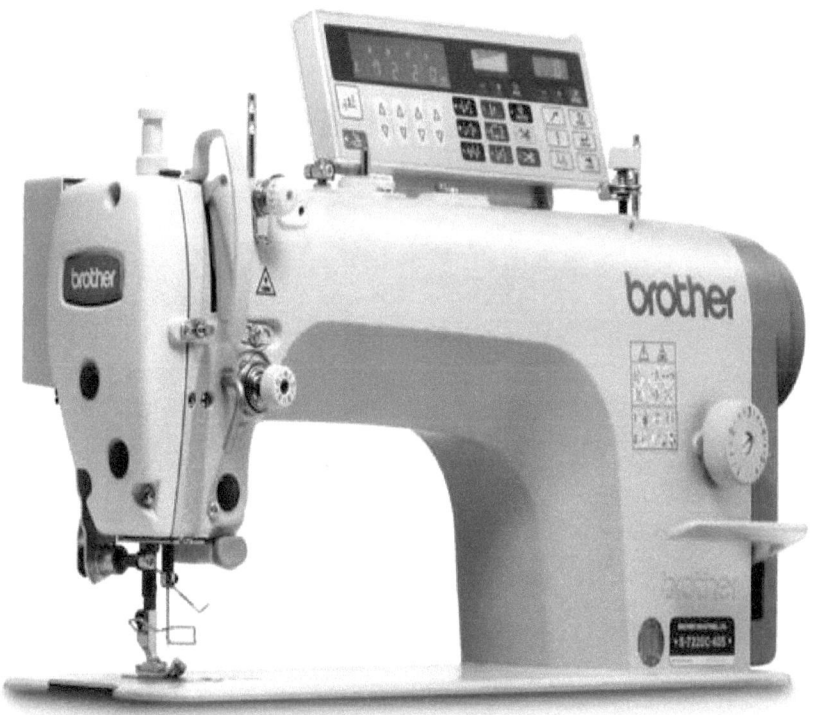

(a) Triple Straight Stitch

This can stretch up to 75% its original length, as it stitches two stitches forward and one back. Used on knitted fabrics for ordinary plain seams or on normal fabrics where the seam is under constant stress, e.g. armholes, sports clothes and trousers.

(b) Overlock

This doubles up as a straight stitching machine and overlock machine by seaming and neatening the seam by overlocking at the same time. The kind of seam made by this machine can be used instead of a French seam.

(c) Super Stretch

This is a highly sophisticated machine capable of making triple stretch and overcasting in one operation. The machine is capable of running without putting pressure on the motor by engaging a slide lever enabling it to run fast or slower. The sewing machine often has electronic incorporated in its name.

Presser Feet that may be used with the Machine

1. Roll Hemmer
2. Gather
3. Hemmer
4. Blind Hemmer
5. Piping Foot
6. Buttonhole

General Care

1. Always keep sewing machine covered when not in use to keep it free from dust.
2. All moving parts oiled must be oiled regularly.
3. Make sure that the plug and flexes are in good condition for electric sewing machine, with no naked wires showing.
4. Tighten all loose screws and bolts.
5. Check that the correct type of needle for the sewing machine is inserted.
6. Always pull the fabric away from sewer after stitching.
7. Never run an engaged threaded sewing machine unless there is fabric under pressure foot.
8. Never turn the wheel to the wrong direction to avoid breaking the needle thread.

Machine Stitching

Check the machine instruction manual for correct threading and its manipulation. Always test your machine on a spare double fabric after threading. This is done to ensure that the correct tension and stitch length is selected.

Correct tension: The stitch should look similar from both sides of the fabric and the thread neither loose nor tight.

Incorrect tension: When the top tension is tight the needle thread will pull tight. Try to loosen the tension adjustment nob.

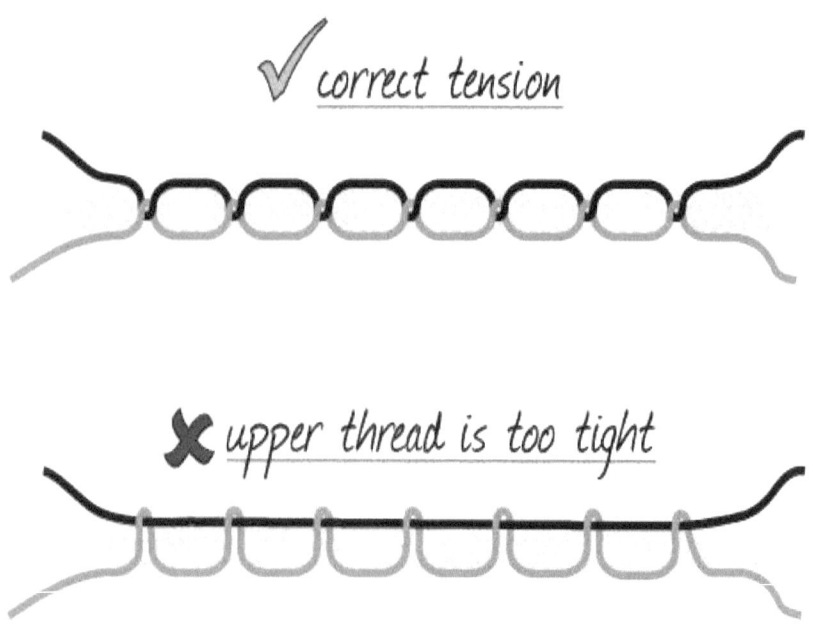

Common machine Faults and Possible Causes

The trainee is expected to understand the common troubles that may be encountered while sewing and must be capable to correct as they are common and bothering and slow down work. A person running the machine ought to be capable to correct the faults and continue sewing.

The cause of uneven stitches

The presser foot holds the cloth firmly. When the pressure of the foot is too light, the cloth is not held firmly in its place so the stitches fall unevenly. You may be pulling the cloth too much. Pressure on the footer may be either too light or too heavy for the fabric. Thread clogging between the teeth of the feed dog might be causing uneven stitches.

The cause of seam-stretching

When the foot-presser is too heavy it does not allow the cloth to move forward, on the other hand the feet-dog is trying to force the cloth forward. So these two opposing actions result into the stretching of the seam. The remedy therefore lies in reducing the foot pressure. When sewing the bias-seams, reduce the foot-pressure.

In each case check the tension first and adjust if necessary.

Persistent incorrect tension: machine threaded incorrect or position of the spool.

Missed stitches: Blunt, bent or needle is wrongly inserted.

Thread breaking: Needle is too fine for the thread, too coarse, or needle is wrongly inserted.

Puckered fabric: Blunt or too large a needle: fabric being pulled through the feed.

Broken needle: needle is bent Fabric pulled towards the front of machine: pins or tacking knots on stitch line: too many thickness of fabric.

1.4 Textile fibers
Fabric fibres

Textile fibers have had long history dating to the early years of 20th century when fabrics could only be produced from natural fibers. Due to increase in the knowledge of chemistry and the extensive research by scientists, man-made fibres have come into being – first, the rayon groups which are chemically processed from a vegetable base, and later, the pure synthetic fibres of a chemical source. All the man-made fibres are processed from a liquid. These new fibres are being constantly improved and varied in their processing to produce not only new fabrics with particular characteristics but fabrics that simulate the appearance and quality of fabrics of a natural fibre; and also yarn and fibres that may be mixed or blended to reduce any disadvantage of natural fibres. Basic fibres are of two main groups: natural and man-made.

General properties of textile fibres

In order to be characterized as a desirable substance to be used as a textile fibre. It must have certain fundamental attributes or characteristics.

1. **High length to width ratio**

Fibrous materials should have sufficient staple length or fibre length. The length should be substantially greater than its diameter. Most fibers have ratios that are much higher. Fibres that are less than ½ inch long are seldom used in yarn production.

2. **Tenacity (adequate strength)**

Although strength of different fibres varies. It is important that the substance possess sufficient strength to be worked or processed by machinery as well as to provide adequate durability in the end use to which it is intended.

3. **Flexibility /pliability**

Flexibility or pliability is the ability of a fibre to crease without breaking. Fibres should be able to bend, pliable or flexible for them to be suitable for yarns. Fibers that are to bend will move with the body and allow freedom of movement, many substances naturally resembles fibrous in forms but due to their stiffness or brittleness they are not used to make textile fibers.

4. **Cohesiveness or spinning quality**

Cohesiveness is the quality of fibers to bind together in yarn making process. The spinning quality of fibres might be due to the lengthwise shape or cross sectional shape that allows them to stick together or entangle while the fibre shape and surface don't contribute to the cohesive quality, similar result may be achieved by using fibres of filament length that are easily twisted into yarn

5. Uniformity

The shape of fibre will include uniform length, surface shape, surface abnormalities and cross-section. Man-made fibres are controlled at the time of production so that a comparatively high degree of uniformity is kept and abnormalities are held to a minimal, yarns that are made of generally uniform fibers are chosen because they are steady, smooth and easy to dye.

6. Elastic recovery and elongation

Elongation is the measure of stretch or extension that a fibre will take before breaking. While elastic recovery is the return from elongation towards the initial length. Some fibres with low elongation have excellent elastic recovery.

7. Resiliency

Resiliency is the quality of a fibre to regain its shape after compression. To establish the crease recovery of a fibre or fabric, elastic recovery of a fibre or fabric is a very important factor in the resiliency of a fibre and normally excellent elastic recovery suggests good resiliency.

1.4.1 Natural fibres

1.4.1.1 Vegetable fibres
Cotton

Source: The cotton fibres surround the seeds of a cotton plant. The fibres are accumulated after the cotton boll or seedpod has split. The length and fineness of the fibres depend on the area in which cotton is grown.

Main region of growth: India, Africa and China. Egypt and West Indies (Sea Island cotton) produce the finest and longest fibres, whereas the United States and Russia produce a shorter tougher fibre.

Production and processing: when the fibres are ready, cotton can be picked by hand or mechanical picking machines or by stripping devices.

Spinning: the spinning mills open and clean the baled cotton fibres can also be blended at this time: the fibres are combed so that they lie parallel a process called carding. Drawing is done in stages until a thin thread is formed. The cotton thread is then twisted and two or more strands of yarns are twisted together.

Weaving: in the weaving mills yarns are coated with starch to strengthen them so that they can withstand weaving process on the weaving looms.

Finishing: woven cloth is inspected to ensure it has no faults. It's then bleached and colour added by dyeing or printing.

Cotton fibres are tiny and uniform in width.

Advantages

1. Cotton fibers are durable thus able to withstand hardwearing even when the fabric is fine
2. Cotton fibres are easy to launder and is strong when wet
3. Cotton fibres conducts heat, thus it's cool to the wearer. Is suitable for summer or sports
4. Absorbs moisture readily and it is therefore good for underwear and children's wear
5. Cotton has good affinity for dyes and chemical finishes are taken easily.

Disadvantages

1. Cotton fabrics tends to crease as the fibers have very little natural resilience
2. Cotton is inflammable and flares up quickly
3. Cotton fibres are weakened by prolonged strong sunlight
4. Cotton will develop mildew if stored damp or in a damp atmosphere

Uses of cotton

a. Table clothes
b. Bed sheets
c. Bath Towels
d. Napkins
e. Cotton wool
f. Bandages

Linen

Source: linen is a basic fibre obtained from the flax plant derived from its stock or stem.

Main region of growth: Belgium, Northern Ireland and the Baltic States.

Linen Production and processing:

The flax plant are pulled up by hand or machines, and are send to the mills where the outer woody portions are rotten away to obtain the fibres. The rotten stems are fibres of the stem are covered with a woody casing comprising gum: to remove the fibres the whole stems are placed in specially constructed ponds and soaked until the long strands are released and separated from the woody casing through a process called scutching. The fibres are then combed to separate short fibres from long fibers in a process called carding. Linen fibers are then drawn out into yarn and twist is imparted.

Linen fibres are composed of smooth, rounded and lustrous bundles of fibers laid together by a gummy substance.

Advantages

1. Linen fibers have high strength to withstand hard wearing even when the fabric is fine

Easy launder because of the good strength when wet.
2. Cool to wear as it will withstand high ironing temperatures so long as the fabric is not held at high temperatures for long time.
3. It absorbs moisture readily and, like cotton, is particularly good for use in hot climates
4. Linen fibres have a high natural luster with attractive yarns and fabrics. The smooth surface of linen makes it dirt resistant.

Disadvantages
1. Linen poor natural resilience thus they wrinkle and crease badly.
2. It is inflammable
3. Linen fabric frays easily.
4. Mildew can develop if stored damp
5. Linen fabrics are expensive since longer grown in large plantations.

Uses of linen
a. Curtains
b. Table clothes
c. Bed sheets
d. Sheer linin fabrics
e. Bath Towels
f. Napkins

1.4.1.2 Animal fibres
Silk

Source: the main source of silk fibres is silkworm cocoon of the breed (Bombyx mori)

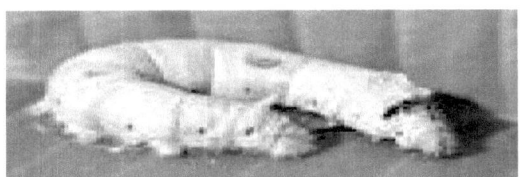

Silk worm (Bombyx mori)

Main regions of growth and production: Japan, China, India and Italy

Silk Production and processing:

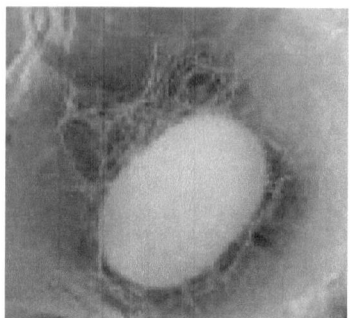

Silkworm cocoon

For about 10 days the worms does nothing but feeds on fresh mulberry leaves until filled with liquid silk. The worm has two small holes in its head known as spinnerets. When ready to spin into a cocoon it bind itself to a twig and starts to spin the two filaments that are exuded from two tiny openings under the head: these continuous filaments are covered with a gum-like substance called sericin. Unless the cocoon is to be kept to hatch for breeding purposes, the pupa is killed by heat before it comes out as a moth so that the cocoon stays undamaged. The cocoons are then put into tanks of heated water to soften the gummy substance and then a rotating brush is used to get the ends of the continuous filaments fibres; several ends are brought together through a guide, they slightly twisted and wound on to a bobbin. Most silk fabrics are woven with the gum still on the threads; this is later removed by boiling the woven fabric. The process of converting raw silk into either one of two types of yarn is called throwing.

Silk fibres under a microscope show their smoothness and lustre.

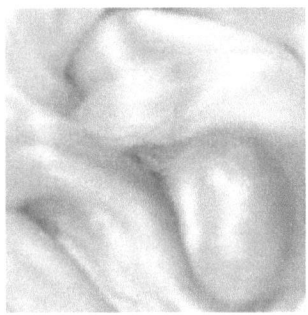

Silk fibers

Advantages

1. Silk has the greatest strength among textile fibres.
2. After removal of gum silk is soft, smooth and lusters.
3. It is warm to wear because it does not conductor heat
4. It absorbs moisture readily which is helpful in application of dyes and finishes
5. The natural resilience and elasticity of the filaments enables silk fabric to drape well and to be crease resistant
6. It is non-inflammable
7. Silk is no damaged by moth.

Disadvantages

1. Silk fabrics are difficult to launder.
2. Silk soils and stains easily.
3. It is expensive

Uses of silk

a. Shirts
b. Ties
c. Blouses
d. Dresses
e. Lining
f. Pajamas
g. Robes
h. Suits

i. Beddings

Wool

Source: The main source of wool is sheep. Several breeds of sheep are raised primarily for wool fibres. Best quality wool is obtained from the merino sheep.

Unprocessed wool fibers

Wool regions producing: Australia, New Zealand, South Africa and Argentina

Wool Production and processing

Grading of wool is done while fibres are still in fleece, graded according to finesse and length. In the manufacture of raw wool into yarn, there are two distinguished processes, one for worsted yarn, and a second for woolen yarn or cloth. Worsted yarns are made from the long and strong fibres only, all the weak short fibres are removed by combing. The long fibres are made to lie straight, paralleled and spun in the way as cotton fabrics to produce a smooth regular yarn which is then woven into a worsted cloth. Woolen yarn is prepared from all the short or broken fibres kept from the combing of the worsted yarn. When spun, the many ends stick out from the yarn, giving it a rough bulk appearance: similarly, all short-fibred fleece and hair is spun for woolen fabrics. A wool fibre under a microscope shows a scale like structure. The size of scale differs from very small ones to large.

Advantages

1. Wool is warm to wear and is a bad conductor of heat: the individual fibres trap the air which acts as an insulation helping to retain the body heat.
2. Wool fibres absorb moisture readily, accepts dyes easily and comfortable to wear.
3. Natural resilience of wool is exceptionally good, it elasticity is excellent as well making the fabric crease-resistant.
4. Wool is water-repellent caused by its natural oils.
5. It is non-flammable but inclines to smolder when approaching to flame.

Disadvantages

1. Wool is Weak when wet, the fibres soften and the fabric will stretch easily.
2. Wool requires careful laundering: should not be hand wrung or agitated.
3. Wool is damaged by several types of insects like moth.

Uses of wool

a. Carpets
b. Socks
c. Sports wear
d. Sweaters
e. Wall hangings

1.4.2 Man-made fibres

Rayon

The principle raw material for rayon is wood pulp and cotton linters. Cotton fibres that are too short for yarn or fabric manufacture are also used. Wood pulp and linters are refined to extract pure cellulose.

Viscose rayon

Production of the filament: the liquid is forced through the tiny holes of a spinneret to eject long continuous filaments which congeal on contact with the air; they are further hardened by treatment with sulphuric acid. The fineness of the filaments is limited by the magnitude of the holes in the spinneret.

Acetate rayon and triacetate rayon

Source: raw material for processing acetate includes cellulose, acetic acid as a catalyst and sulphuric acid.

Production of the filament: the flakes are dissolved in acetone and the sticky liquid produced is forced through spinneret and ejected into a current of warm air to solidify the filaments.

Cuprammonium rayon

Source: from cotton linters treated with copper sulphate, caustic soda and ammonia.

Production of the filament: the liquid is forced through a spinneret and the filaments hardened in a chemical tank; they also become more elastic and can be drawn out to finer degrees of thinness.

Filament yarn: the continuous filaments are spun together and woven to produce smooth fabrics such as satin and taffeta.

Staple yarn: the filaments are cut into shorter lengths and spun to produce a more bulky yarn for heavier weight fabrics.

Rayon fibers

Advantages

1. Resemblance to silk fabrics but less expensive
2. Viscose is warm to wear
3. Viscose is absorbent and dries quickly
4. Viscose does not crease easily.
5. Viscos is resistant to insects like moth proof and resistant to mildew.

Disadvantages

1. Requires careful laundering as the fibres are easily damaged by rough or heavy handling
2. Is easily damaged by chemicals such as acids and bleaching agents.
3. Can be damaged by excessive heat, dry or moist.
4. Rayon is inflammable

Uses of viscose
- Apparel
- Home finishing
- Automobile tires

Nylon
Production of nylon

Nylon is made from two basic chemicals a dipic acid and hexamethyline diamine. Nylon is made by liner condensation. When a molecule of acid and of diamine join together a molecule of water is given off. The elements found in nylon are also found in cool air and water. The chips are then melted and the liquid forced through a spinneret to form filaments which solidify on contact with the air; these filaments are smooth and rod-shaped.

Nylon fibers

Advantages

1. Nylon is a very strong fibres
2. Nylon fibre has excellent elongation properties and is very elastic.
3. The smooth surface of the filament repels loose dirt to a certain degree and is easy to launder.
4. The luster of nylon can be controlled and may vary from dull to bright.
5. Nylon dries quickly after laundering as a result of low moisture absorption.
6. Nylon has good resiliency resulting to good recovery from creasing.
7. Nylon is moth proof.
8. Is inflammable.

Disadvantages

1. Frays easily
2. Nylon Can be damaged by some acids and bleaching agents but is not affected by alkaline fluids

3. Nylon has low moisture absorption therefore not suitable for young children's wear and for sportswear for use in summer.

Uses of nylon
1. Carpets
2. Dresses
3. Gowns
4. Blouses
5. Shirts
6. Handbags
7. Hats curtains

Polyester

Polyester is the product of the reaction between ethylene glycol and terephthalic acid (both from petroleum)

Production of filament yarns: the basic substances are processed to produce an ivory coloured plastic which when solidified is cut up into chips: these polymer chips are melted at a high temperature and forced through a spinneret. The polyester yarn is then produced in two distinct forms:

Filament fibre: as the filaments are ejected from the spinneret, and solidify, they are wound on to cones as 'undrawn' yarn; this yarn is then drawn and stretched for a number of times to its original length and then wound on to bobbins.

Staple fibre: as the filaments come from the spinneret they are brought together into a thick bundle called a 'tow' and then drawn out. The 'tow' is then artificially waved and set by heat, chopped into specific lengths and then spun in the usual manner.

Polyester fibres

Advantages

1. Polyester is a strong fibre but with slightly less strength when wet
2. Polyester is easy to launder and dries quickly; it requires little ironing
3. Polyester is crease-resistant but can be heat set for permanent pleating.
4. Polyester is smooth, soft and drapes well.
5. Dry cleaning agents, alkaline or acid substances when used in moderation do not damaged polyester
6. It has a high resistance to damage by light and sunlight through glass and is therefore ideal for curtains
7. Is both moth proof and resistant to mildew

Disadvantages

1. Polyester fibres are not absorbent unless mixed with other fibres
2. The fabric frays easily

Uses of polyester

1. Laundry bags
2. Bed sheets
3. Pillows
4. Upholstery
5. Carpets and curtains
6. Dresses

Acrylic fibres

Acrylic fibers are polymers formed by addition polymerization of chemical acrylonitrile and dymithyle formamide solvents. The dissolved polymer is extruded through a spinneret into heated spinning container where filaments solidify. The filaments are stretched while still hot to introduce molecular orientation.

Acrylic fibres

Advantages

1. Acrylic fibers are strong and are resistant to abrasion.
2. It is warm to wear as the air is trapped in the staple fibre

3. It is easy to launder and has good wash and ware properties.
4. Drapes well
5. Acrylic fibers have good resiliency thus they will resist wrinkling.

Disadvantage

Static electricity will build up particularly when the humidity is low

Uses of acrylic
1. Blankets
2. Carpets
3. Special floor coverings
4. Ski clothes
5. Snow suits
6. Sports shirts
7. Sweaters

Blends and Mixtures

Fibres are mixed or blended to obtain the advantage of each fibre and to counteract the disadvantages, e.g. polyester and linen. Cotton has the main disadvantages in that it creases badly and has no luster; polyester is crease-resistant therefore, by mixing, the fabric is less crease-resistant than polyester but is more resistant than cotton. The mixing or blending of fibres can reduce the cost of the fabric, e.g. silk mixed with a fine woolen yarn: as the fibres are similar in construction the characteristics are little changed but the cost is lower than if the fabric was purely made of silk. The properties of mixed or blended fibres depend on the proportion of the fibres used: the fabric should be treated in the same way as fabrics made from the fibre of highest proportion.

It is important to choose the correct fabric for your garment. Choosing a fabric which is unsuitable for its purpose will mean that the garment will never achieve the intended quality.

The most important points to take into consideration when choosing a fabric from the colour and pattern are

1. *Weight*. Choose a fabric suitable for the time of year and for the purpose of the garment.
2. *Laundering*. Remember that clothes in regular use should be made of a fabric which can easily be laundered.
3. *Content of fibre*. Find out the percentage of the main fibres to obtain the properties of the fabric.
4. *Texture*. Choose soft plain fabrics, perhaps with a slight sheen, for intricate draped styles but use more highly textured bulky fabrics for outdoor wear.

1.5 Revision Questions

1. The following are ways in which you can turn dressmaking into money. Which one is not?
    A. Producing patterns for sale
    B. Being employed to sew or as a machinist
    C. Being a secretary
    D. Open a garment making establishment of your own
2. Select the types of garments where fly openings are applied on.
    A. Sweaters
    B. Trousers
    C. Overcoats
3. Which statement does not explain the use of tissue paper in a dressing workshop?
    A. Care of white fabrics and those which crease easily
    B. Wipe excess oil on the sewing machines
    C. Support seams which are to be machine stitched on very fine fabrics such as chiffon
4. Which one of the following list of tools is not used for marking?
    A. Tailor's chalk
    B. Tracing wheel

C. Thimble
5. Which one of the following is not the correct use of pins in dressmaking
    A. Fixing paper patterns on cloth before cutting
    B. Pins are used to hold layers of fabric together before stitching
    C. Piercing your ears
    D. Pins are used to tuck up looseness while trying on
6. Which of the following is not used during pressing
    A. Iron box
    B. Iron board/ pressing board
    C. Sleeve board
    D. Roller
    E. Cotton wool
    F. Pad
    G. Damp cloth
7. Which one of the following statements is not true about caring for silk fabric
    A. Moderate iron on the wrong side
    B. Shrink when washed
    C. Do not use damp clothes because they may leave water marks
    D. Do not use steam iron because it might leave water marks
8. Which one is not a way of caring for a sewing machine
    A. Check for loose screws
    B. Leaving your machine uncovered while not in use
    C. Keep the moving parts oiled
    D. Use the correct type of needle for the sewing machine
9. Which one of following parts of the sewing machine engages and disengage the stitching mechanism of a sewing machine
    A. Needle clamp
    B. Thumb screw
    C. Feed dog

D. Hand wheel
10. The following are groups of sewing machines, which one is not?
    A. Straight stitching machines
    B. Semi-automatic machines
    C. Automatics and Superautomatics
    D. Darners
11. Which one of the following is not a natural fibre.
    A. Cotton
    B. Silk
    C. Wool
    D. Rayon
12. Choose the advantage of linen from following list that is not similar to cotton
    A. Hard wearing even when the fabric is fine
    B. Launders well
    C. Linen is shiny
    D. Absorbs moisture readily
13. Which one of the following is not a disadvantage of silk
    A. Difficult to launder
    B. Stains and watermarks easily
    C. It is expensive
    D. Its elastic
14. Which of the following statements is true about worsted yarns?
    A. Prepared from short yarns
    B. Prepared from short fibres
    C. Prepared from long and strong fibres
    D. Prepared from a blend of fibres from different breeds of sheep
15. Which one of the following is not a disadvantage of Nylon
    A. Frays easily
    B. Damaged by some acids and bleaching agents
    C. Is non-absorbent
    D. Is inflammable

16. The following are reason for blending or mixing fibres, which one is not?
    A. Obtain the advantage of each fibre and counteract the disadvantages
    B. To change from their colour
    C. Reduce the cost of the fabric
17. Which one is not measures to consider when choosing fabric for a garment
    A. Weight of the fabric
    B. Laundering
    C. Fibre content
    D. Country of origin
18. Which one of the following is not a possible causes of fire outbreak in a workshop
    A. Electric fault
    B. Sewing machine catching fire because of being overworked
    C. Smoking in the workshop
    D. Careless handling of pressing equipment

Supplementary Content

1. Personal Appearance

A person will gain confidence when they are well groomed. This will be achieved by careful selection of the clothing to dress on, putting into consideration the colour, style and fabric in order that they may enhance the good features while dispensing, or at least minimizing, those which are less attractive. Details adaptable to your own style and personality should be considered while one is striving follow fashion trends.

Budgeting

When it has been decided how much money to be used on clothes each year. It is best to plan for the major items first. However the full scope of the budget must be kept in mind, which will include the whole wardrobe and its maintenance.

Allowance must be made for all accessories, dry cleaning and shoes repairs in addition to top clothes which will include casual and sportswear, summer and winter wear. Also the requisite underwear for the top clothe, night wear, sock, stockings tight and foot wear,

Remember that in some regions climate can seldom be trusted to provide long spell of warm weather so allocate money for summer dresses accordingly and do not be tempted in the spring to buy several of new seasons' lightweight dresses, which at the end of a poor summer may be scarcely worn and be outdated by next summer season.

Coasts and suits are the main items in the budget; choose a style that is not of high fashion so that it will not be out of date by the next year. It is wise to buy the best that you can afford as a good coat or suit should at list last for two seasons and poor quality clothes soon appear shabby and become out of shape. When planning new purchases remember the clothes that you have so that the new clothes can integrate with them in use and widen the possibilities of different combinations.

Whenever possible, try on the Ready-made clothes to make sure that it both suits you and fits well. Thereby you may void an expensive mistake.

Store sales – by buying carefully you may be able to extend your budget, look for well-known trade names of fabrics and clothes which have been genuinely reduced. Beware of cheap quality goods that may have been 'bought in' for the sales.

Avoid buying goods in a hurry without proper consideration, take time to select and if undecided leave the problem and return when a proper decision has been made.

Colour

The choice of colour is most important as it should enhance the looks of the wearer, her personality and the style of the garment. Study the colour of your eyes, hair and complexion, always try to compliment the eyes which are the focal point of your expression.

Those with fair colouring can wear most colours well, in particular the soft shades. Vivid colours can best be worn by those with dark hair or sallow skins so these provide a definite contrast. People with high complexions, red or auburn hair often look best in cool colours such as green or blue.

Dark colours have a slimming effect whereas pale colours which have more impact, have an enlarging effect.

Be adventurous whichever colour is in fashion, one of its shades will suit you even if only a small proportion is worn such as a scarf. Ensure however that it will combine with your existing clothes – often unusual but interesting and satisfactory combinations can be achieved this way.

Style and Choice

The type of figure that you have, its proportion and characteristics should be considered when selecting the style of a garment. Determine which are good and bad points of your figure and look for a style which will disguise the bad ones.

To disguise the faults naturally gives emphasis to the better points of a figure. It is foolish to choose a style which you will find very attractive but does not suit you.

It is necessary to consider the purpose of the garment and the practicality of the style. It is also important to consider the colour and texture of the fabric to be used.

Tall and Angular Figure

Choose:

1. Contrasting colour that would seem to break the height.
2. Double-breasted coats and jackets
3. Fabric with bright and large patterns
4. Dresses that fall from the shoulder or yoke lines
5. fashioned trousers

Avoid:

1. Tight fitting garments
2. fabrics that hold on tightly
3. Single coloured Outfits

Short and Slim Figure

Choose:

1. Clothes that draw attention to the waistline.
2. Skirts pleated into waist
3. Soft colour and one colour outfits
4. Small printed patterns

Avoid:

1. Heavy and bulky fabrics
2. Large prints and patterns
3. Large tight belts
4. Colour breaking outfits

Tall Plump Figure

Choose:

1. Simple style lines
2. Attractive collars or neck line
3. Separates that by-pass the waist lines
4. Fabrics that are not shiny

Avoid:

1. Clothes with complex style lines
2. Sleeves with fullness, e.g. pleats or gathers at sleeve head
3. Fabrics that cling or have shiny surfaces
4. Large fabric patterns
5. Light colours

Short Plump Figure

Choose:

1. Simple styles with vertical seams.
2. Fabrics with vertical stripes or small patterns
3. Straight panel dresses.
4. 'V' neck lines

Avoid:

1. Anything tight or waisted
2. Fabrics which are either bulky or with large bright patterns
3. Trousers
4. Separates that contrast too much in colour
5. Frills

Figure with a Large Bust

Choose:

1. 'V' neck lines
2. One-piece dress
3. Separates with a dark top and light skin

Avoid:

1. High neck lines
2. Empire line dress
3. Frills on the bodice
4. Tight belts and tightly fitting waist
5. Short sleeves

Figure with Large Hips

Choose:

1. Styles that hand from the shoulder or a chest yoke line
2. Interesting neck line detail
3. Separates with a bright top and dark-coloured skirt

Avoid:

1. Trousers
2. Hipster skirts
3. Shirt or full-bottomed sleeves
4. Tight belts at waist or hip level

2. Fabrics for everyday garments

C: Cotton W: Wool S: Silk P: Polyester R: Rayon N: Polyamides (Nylon group) A: Acrylic

Garment	Lightweight		Medium		Heavyweight	
Blouse	Cheesecloth	C	Calico	C, P	Sailcloth	C, P
Shirts	Lawn	C, P	Cambric	C	Fine wool Angora	W
Tops	Georgette	C, S, P	Crepe du Chine	S	Wool Jersey	W
	Muslin	C, P	Dimity	C		
	Nainsook	C	Gingham	C, P		
	Organdie	C, P	Pique	C, P		
	Voile	C, P	Poplin	C, P		
			Seersucker	C, P, N		
			Tussah	S		
			Viyella	W		
Beachwear	Cheesecloth	C	Gingham	C, P	Terry Towelling	C
	Lawn	C, P	Helenca	P		
	Jersey	C, P	Poplin	C		
			Stretch Towelling	P, N		
Nightwear	Lawn	C, P	Brushed Rayon	R	Flannelette, Wincyette / Flame proof	C
	Jersey	C, P, N	Dimity	C		
			Poplin	C		
			Seersucker	C, N		
Dresses	Cheesecloth	C	Brocade	S, R	Angora	W
	Lawn	C, P	Brushed Fabric	C, P, R	Boucle	W, S
	Nainsook	C	Corduroy	C, P	Flannel	W
	Poplin	C, P	Courtelle	P	Grograin	S, R
	Sarille	R	Crimplene	P	Tweed	W, P
	Tricel	R	Jersey	C, S,	Velvet	S, P

					W, P, N		
	Tussah	S	Orlon	A			
			Sailcloth	C, P			
			Velvet	C, P, R			
			Velveteen	C, P			
			Viyella	W			
Jackets	Courtelle	A	Boucle	W	Barathea	W	
Skirts	Jersey	C, W, P, N	Corduroy	C, P	Cavalry Twill	W, N, P	
Trousers	Orlon	A	Denim	C, P	Donegal Tweed	W	
	Pann Velvet	P	Drill	C	Drill	C, W	
	Treveira	P	Sailcloth	C	Flannel	W, P	
			Velvet	C, P, R	Fur	A	
			Velveteen	C, P	Gaberdine	W, P	
			Velour	C			

(This table is extracted from Valerie I. Cock. 1999. *Dressmaking Simplified*. Third Edition. Malden, MA, USA: Blackwell Science, Inc.)

Chapter 2: Garment Construction

2.1 Taking measurements

It is necessary to be accurate when taking measurements, check that the tape measure is correctly placed to obtain the correct body measurements.

All round measurements require added allowance for ease of movement – as given below.

Before taking the measurements determine the waist position by tying a piece of tape firmly but comfortably around the waist.

Do not pull the tape measure too tightly.

2.1.2 How to take direct measures

   a. **Shoulder**

Shoulder measurement is taken from the sleeve on one shoulder to the other with the tape pissing on the nape.

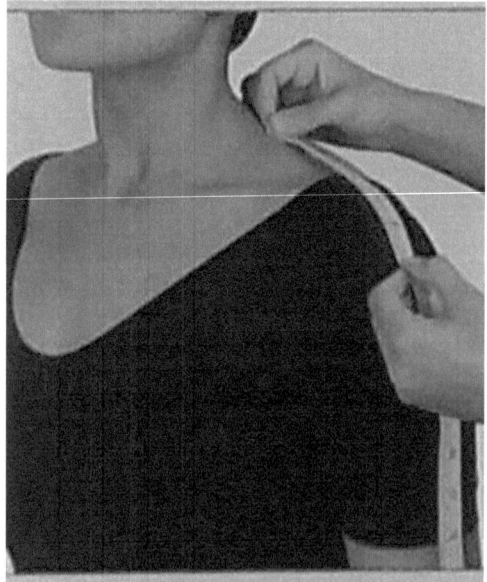

   b. **Bust /chest**

It's taken around the chest with two fingers between the bust and the tape. The tape should be held horizontally on the fullest (prominent) part of the bust.

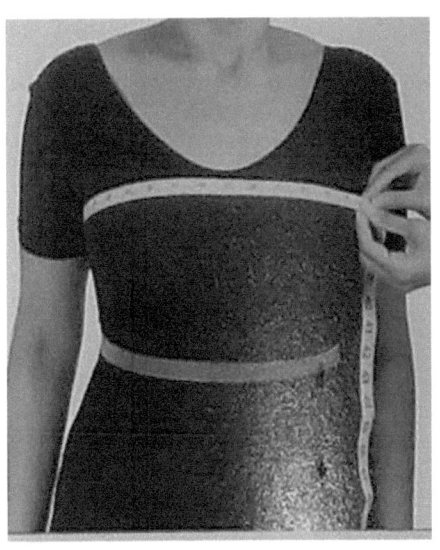

c. Full length for shirt or blouse

This is taken from the top of the shoulder close to the neck down to the desired length.

d. Sleeve length

Sleeve length is measured from the sleeve seam on the edge of the shoulder down to the desired length.

e. Waist length

Measured from the nape down to the hollowest part of the waist.

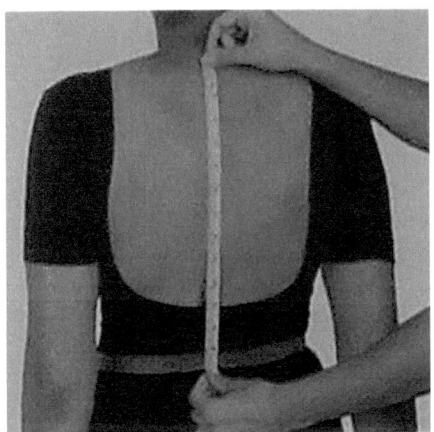

f. Waist measurement

Measure the hollowest part of the waist, is take over the skirt, dress or trouser.

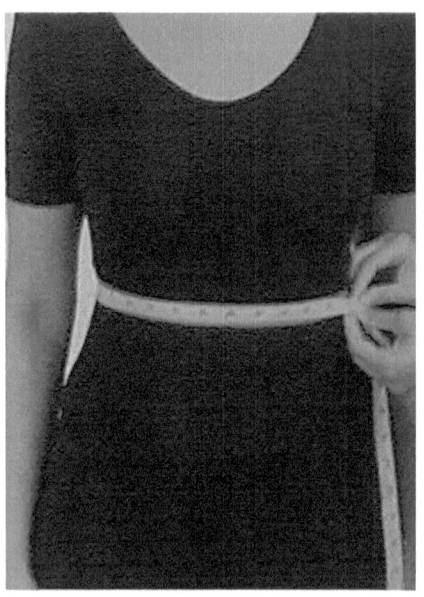

g. Seat or hip measurement

Measure horizontally around the most prominent part of the seat.

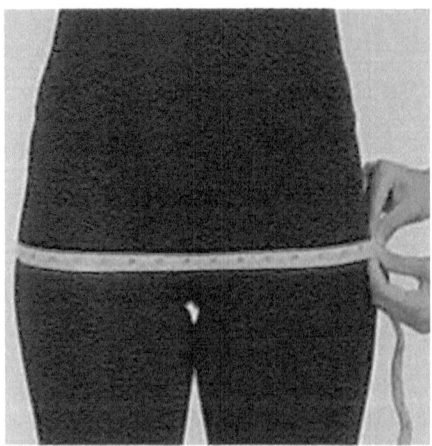

h. Full length for trouser or skirt

Measured from 2" above the hollowest waist down to the desired length for skirts, trousers are measured to the floor.

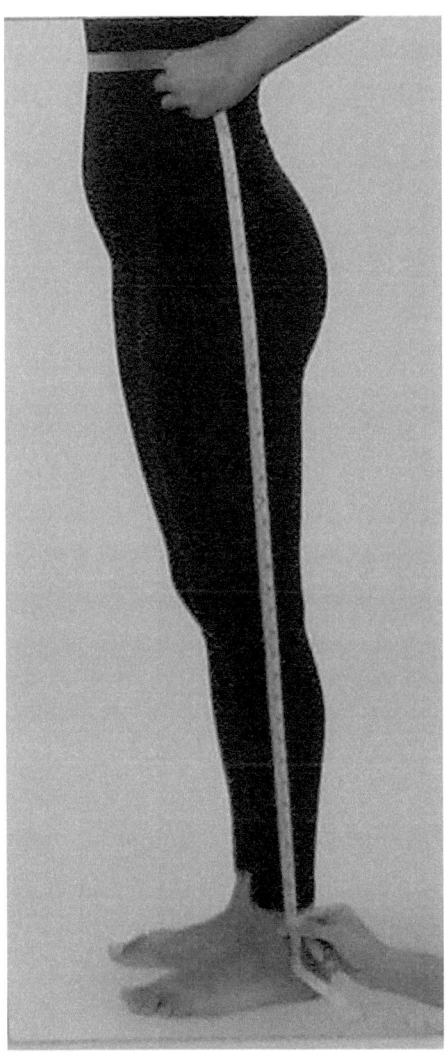

i. Knee width

Is measured around the knee as per one's preference.

j. Bottom width

Measured all round above the sole to the customers preference.

2.2 Garment design

Before you designing your own garments it's important to recognize good designs and what attracts you when you see it on another person. Being able to choose features of styling and discover the reason you like a particular garment will enable you to get inspiration from those particulars and apply them in your designs.

There are particular features of design, that both applicable and reliable, that should be thought out when you plan for a garment.

1. **Purpose of your dress:** when and where do you intend to put on your dress? For what nature of events or activities? Is it business or casual?
2. **Dress styles:**
   Dresses have the main item of female apparel in for centuries. The three classic dresses are
   Sheath: a close-fitting dress that is shaped by darts.
   Shift: a loose fitting dress.
   Princess: a close fitting flared dress that is shaped by seams.

Neckline styles

A neckline refers to the region surrounding the neck and shoulders. Fashion features many different types of necklines in a season.

Collar styles

A collar is a distinguished piece of cloth that is sewn to the neckline of a garment. It can be large or small, stand up or fold-over.

Classic collar styles:

   a) Shirt collar
   b) Peter pan collar
   c) Convertible collar
   d) Shawl collar
   e) Roll collar
   f) Mandarin collar

Sleeve styles

A sleeve is the apart of a garment that is sewn at the armhole to neaten the armhole and cover to the arm.

a) *Set in sleeve:* is attached to the garment by the armhole seam, the sleeve is set into the armhole when the shoulder and the underarm seams have been stitched.
b) *Raglan sleeve:* the sleeve extends to the neckline with front and back diagonal seams that extends to the neckline from the armhole.
c) *Kimono sleeve:* is cut together with both the front and the back patterns.

Shirt style:

The term shirt is unusually used to describe a piece of clothing that is more tailored than a blouse.

Skirt styles

Skirt is described as a separate piece of garment that can be dressed with any fashion of top. Skirts can be straight, flared or full darts, gathers, pleats or seams shape them.

Pants

An outer garment worn from the waist downwards and separates to each leg worn by men, boys, women and girls. Pants styles vary in width and length.

Jacket and coats style:

Jacket can be worn as an outer layer or under coat. A coat is a garment that is worn on top of the others and has sleeves and covers the body from shoulder down and worn outdoors. Jackets and coats can be single breasted or double breasted, long or short.

2.3 Pattern alteration

It is often necessary to make small adjustments to a bought paper pattern in order that the main measurements may correspond with your personal proportions. This would make garment to fit appropriately and look attractive.

2.3.1 To Reduce the Measurements of a Pattern

At the most suitable point, fold the paper to make a pleat whose width is equal to half of the amount to be reduced and pin in place. To adjust the width, the extra amount is divided between the back and front and evenly between left and right sides of the garment.

To correct the seam lines it may be necessary to add a strip of paper and to re-draw.

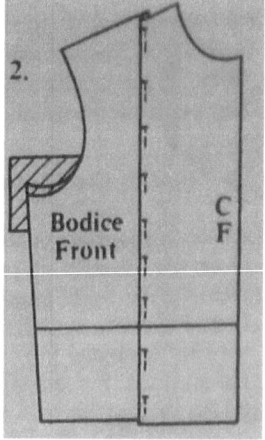

a) Pin Pleats at the back and front bodice to reduce the width of the shoulder, bust and waistline
b) Mark line between bust and waistline to reduce length from neck to waistline
c) Pin a piece of paper behind the pattern the draw a line that raises the under-arm

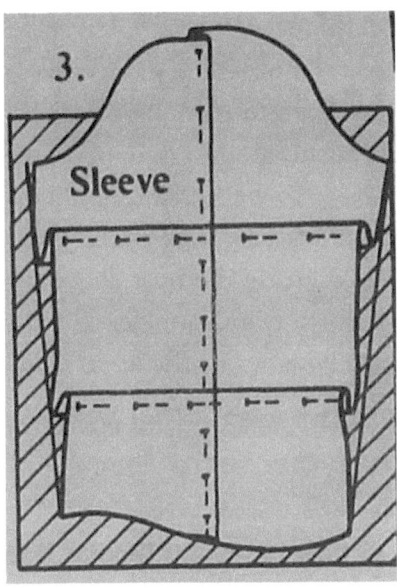

a) To adjust the sleeve width reduced from shoulder to the wrist
b) Adjust sleeve length by reducing between shoulder and elbow, elbow and wrist
c) Rectify the seam line

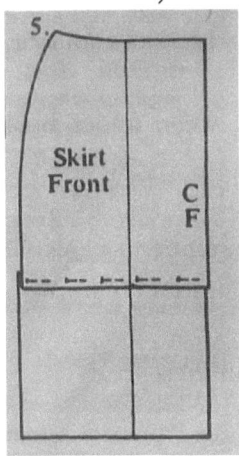

To shorten a skirt – pleat between hips and hemline
Draw the line for reducing the width of waist, hip and hemline

2.3.2 To Enlarge the Measurements of a Pattern

At the suitable line cut through the pattern and place the pieces over a strip of paper so that the edges of the two pattern pieces are parallel, the distance between them will be equivalent to the amount to be added, while altering the width the amount to be added should be divided between back and front patterns and evenly between left and right sides of the garment. For instance to increase the bust and waist measurements by 5 cm insert 1.25 cm cut down from shoulder to waist in all quarters of the bodice. Trace the seam lines across the insertions.

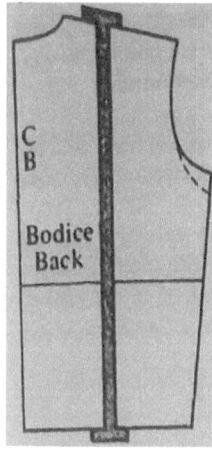

a) Insert a strip of paper to increase width at shoulder, bust, and waistline
b) To increasing length draw a line between under bust and waist
c)
d) The dotted line shows the area to be snipped to enlarge the armhole, ensure that the same amount cut from back and front patterns.

NOTE: the sleeve head must be deepened if the armhole is enlarged.

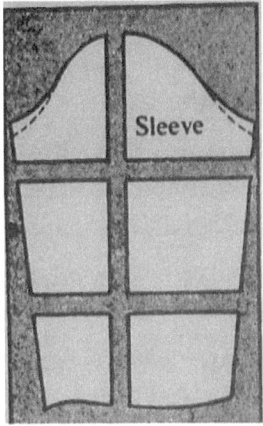

a) Increase width of sleeve from shoulder to the wrist
b) To lengthen a sleeve add between shoulder and elbow, elbow and wrist
c) Drawn the seam lines again and deepen the sleeve head

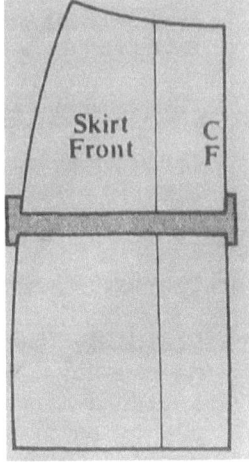

a) Lengthen a skirt between hip and hemline
b) Draw the line for increasing width on waist, hip and hemline

NOTE: After altering the waistline both bodice and skirt must be altered as well

2.3.3 Use of a Paper Pattern

Having selected the pattern buy the necessary amount of fabric and lining, usually two and more versions of the style are given, requiring different amounts of fabric. Buy also the recommended size of buttons, zip and amount of trimmings.

First, open the pattern and study the markings together with the instruction sheets so that they may be fully understood.

Select the pieces of pattern required for your style and find the layout that is correct for both size and style and the width and type of fabric purchased. Replace the pieces not required in the packet.

2.3.4 Preparation of the fabric

1. **Plain fabrics and those with a woven pattern** – straighten the cut ends by drawing a thread across the width and trimming back to this line.

   Printed fabrics - cut the ends straight with the pattern. Be cautious when purchasing printed fabrics as they can be inaccurately printed, sometimes as much as 15 to 16.5 cm off the thread, in which case DO NOT BUY as if the pattern is followed the garment will hang badly because of the bias cutting or if the thread is followed the pattern will be seriously out of place.
2. If fabric is creased press carefully.
3. Shrink loosely woven woolen fabrics by pressing under a damp cloth and allowing to dry in the air
4. Check the fabric for flaws, mark their position so that they may be avoided if possible
5. Examine the fabric and note if it has a pile or nap, as these fabrics require particular attention when laying on the pattern. Usually specific diagrams are given for these fabrics if they are suitable for the style.

2.3.5 General layout of a pattern

1. Fold the fabric as directed and spread squarely and smoothly on the cutting out table

2. Lay out all the pattern pieces required to ensure that there is sufficient fabric
3. Check the position of each piece, so that the necessary pieces are placed to the fold and that the necessary pieces are placed to the fold and that the straight threads (or print) are parallel with the markings for the straight thread
4. Pin the pattern in place with the pins near the seam allowance and at right angles to the edges
5. The fabric must be kept flat on the table and each piece smoothed out as it is pinned.

One-way Fabrics

Since one-way fabrics do not allow patterns to be laid in the opposite direction. Extra yardage is usually necessary when using fabrics with:

One way printed fabric

(a) One-way print
(b) One-way arrangement of woven stripes
(c) With nap – woolen and mixture fabrics with surface fibres lying in one direction only. These should be cut so that the fibres run down towards the hem.
(d) With a cut pile – silk velvets with the pile lying upwards and velveteens, needlecords and corduroy with the pile running

downwards – made up in this way the fabric has a richer colour.

It is essential that all pattern pieces are laid in the same direction, main sections can rarely be interlocked for economy of fabric.

Fabric may be folded in half lengthwise but must not be folded across the width as this will reverse the pattern on pile.

When laying out on woven stripes, one-way prints and prints with a definite repeat motif also follow the instructions for check and striped fabric.

Checks and Striped Fabrics

For a checked garment to be attractive it will call for a professional finish were the checks and stripes match in line at the shoulder, under-arm and side seams, centre back and centre front lines, hemline and all seams of the skirt. Therefore extra fabric must be allowed.

Checked fabric

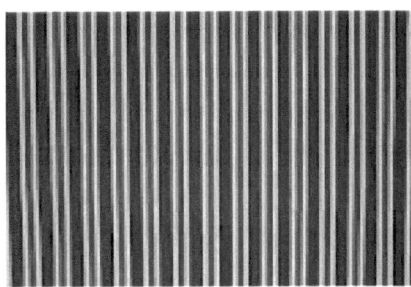

Striped fabric

1. Ensure that fabric is folded correctly so that all lines are exactly in position
2. Place the pieces to be laid against the fold first and position the other pieces accordingly
3. Place the balance marks, centre and hem lines of the sections to be joined in identical position on the checks or stripes

2.3.6 Marking out the Pattern

Tailor tacking is the best method of transferring the pattern markings on to the double fabric as both layers can be accurately marked together. Separate coloured threads must be used to mark different pattern details to make them easy to identify. The colours of thread should be used on balance marks, darts, fitting points, position of buttonholes, centre lines and straight tacking on pleats. Mark with straight tacking the fitting line through single fabric of all curved and detailed sections and small pieces following the fitting lines across seam allowance.

Alternative methods

Tailors' chalk – this can be used if the garment is to be tacked together and made up directly after cutting out, as it will brush off with handling. It cannot however mark both layers at the same time so these must be parted and marked separately, therefore ensure that sleeves are paired and right and left sides marked correctly.

Carbon paper – this may be used with care on washable fabrics only, otherwise possible alterations may be visible. Place the carbon paper between the fabric and the pattern with carbon (shiny) side against the fabric, transfer the markings with a pencil. Double layers have to be marked separately as with tailors' chalk.

2.3.7 Preparation for Fitting

After transferring the pattern markings, remove the main pieces of paper pattern and separate the double or folded pieces, cutting through the tailors' tacks. Run in gathering threads where necessary, prepare tucked areas of fabric, put in gathering threads for any smocking.

Tack together the main sections of the garment in the following order:

1. Darts and pleats
2. Panel lines, shoulder and under-arm seams of the bodice (not the collar or facing)
3. All sections of the skirt
4. Tack the skirt to the bodice and ensure that the position of the opening is correct
5. Tack together a sleeve but do not insert

When preparing a skirt for fitting the best results are obtained if the waistband is fully made up, with the interfacing and tacking it into place on the skirt. This enables the position and measurement of the waistline to be checked thoroughly.

When tacking darts and seams first match and pin together all balance marks and fitting lines carefully. Accuracy at this stage is most important. Place the pins at right angles to the fitting line.

Fitting the Garment

Try the garment on right sides out and note the following points:

Mark any alteration on one side only and transfer the alterations to the other side after the garment has been taken off.

1. Check the hang of the garment on both back and front, ensure that the centre lines are on their correct position and perpendicular.
2. Ensure a garment that has a waistline has it corresponding with the natural waistline of the wearer.
3. Check if there is looseness across the back of the shoulder, snip the tacking and raise the back bodice and re-pin the shoulder seam. Taken up at the shoulder point slopping shoulders
4. Extra fabric should be taken in to form a small dart each side of the centre back
5. Check the position of the bust line and fit of the bust darts, ensure that the darts are placed on the correct position and required size.
6. Width of the garment, check if there are any visible lines across the garment to denote tightness. Therefore release the tacking at the side seams to give a looser fit with a smooth seam line. If width is too loose re-pin side seams to give tighter fit.
7. Ensure that all waist darts correspond on skirt and bodice
8. Pin the sleeve into the armhole to achieve the correct hang and position of ease to give a smooth finish without tight wrinkles or fullness.
9. check neckline, position of fastenings and any trimmings

To Transfer the Alterations

After all the alterations have been noted and marked. The garment should be removed carefully. Tack all the alterations with a new tackline on each section. Following the line of pins tack in the new fitting lines to machine stitch the details and seams.

The pins are placed along the fitting line when fitting and marking alterations.

2.4 Garment construction procedure

2.4.1 Order of Work

1. Preparation of fabric by ensuring that the fabric is free from creases
2. Mark the pattern on the fabric
3. Cut out the patterns
4. Transfer pattern marks
5. Run the gathering threads and tack pleats in position
6. Tack together basic sections of garment for fitting if necessary.
7. Fitting and alteration if necessary
8. Transfer alterations to the other side of garment.
9. Disposal of fullness – darts, pleats, tucks or smocking
10. Shoulder seams and panel lines
11. Make and attach pockets,
12. Collar
13. Front and neck facings,
14. Side seams and skirt seams
15. Wrist opening, sleeve seam, cuff and/ or sleeve edge neatening
16. Insert sleeve, face armhole if no sleeves
17. Join skirt to the bodice
18. Plackets and zip fasteners
19. prepare buttonholes
20. Hemline after final fitting
21. Final press
22. Attach buttons, hooks, eyes and press studs

2.4.2 Stitches

2.4.2.1 Temporary Stitches

Temporary stitches are made to hold two or more layers fabric temporary during stitching, fitting and checking dart positions. These stitches are removed after stitching the garment. These are of various types.

Tailor Tacking

Used for marking through double fabric, the position of darts, buttonholes and other parts of construction.

Using a double thread, work double stitches, leaving long loops, then gently pull the two layers apart and cut through the stitches, leaving half of the thread in each layer of fabric.

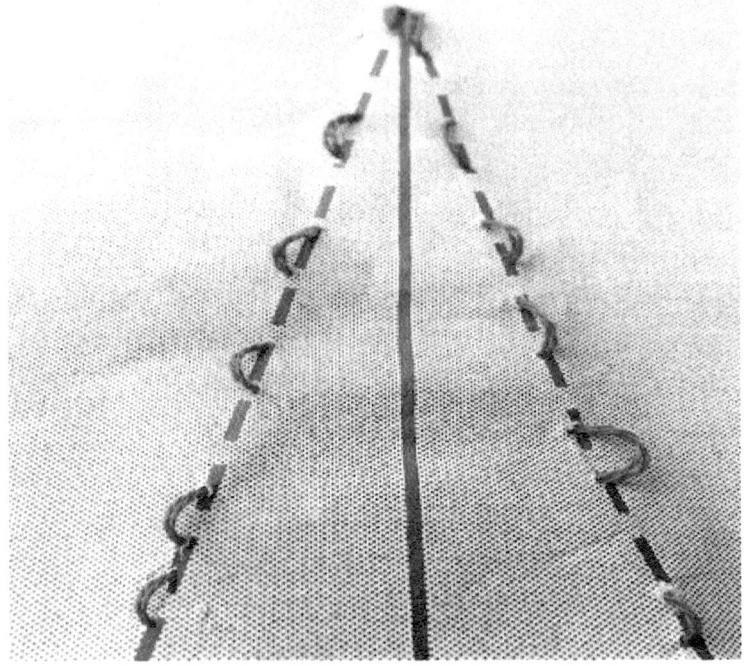

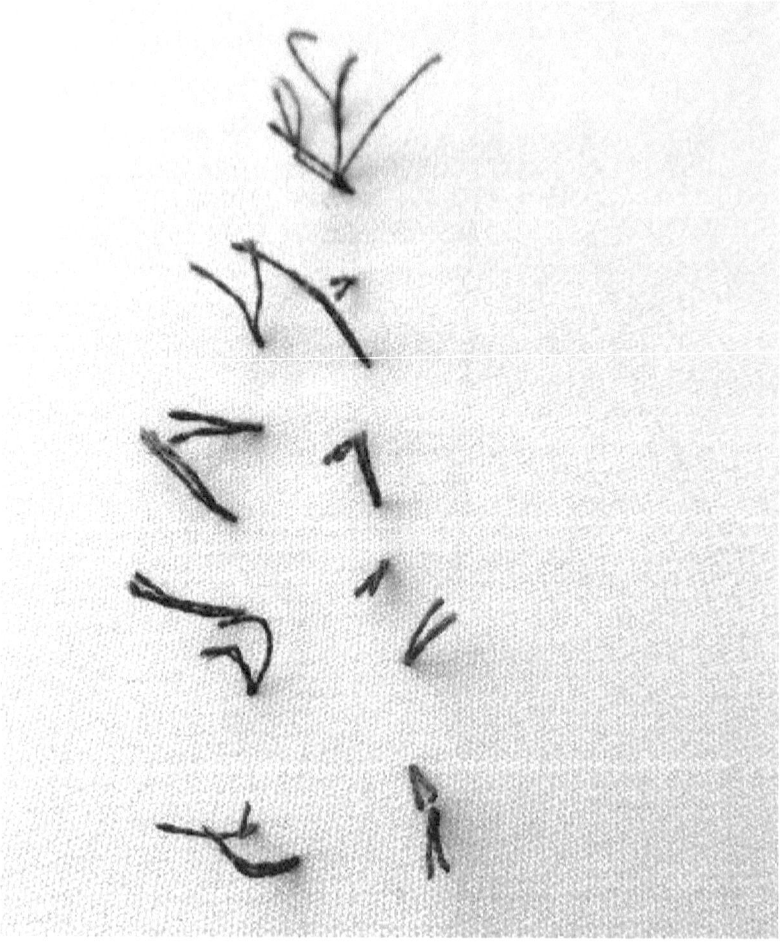

Upright Tacking or Basting
Upright Tacking (also known as Basting or Diagonal Basting)

Use: for holding in place pleats, revers, interfacings and other areas of double fabric during the construction of a garment

Tacking (also known as Straight or Flat Tacking, Basting and Thread Marking)

Use: for marking fitting lines and alteration on single fabric and for holding layers of fabric in position for final stitching.

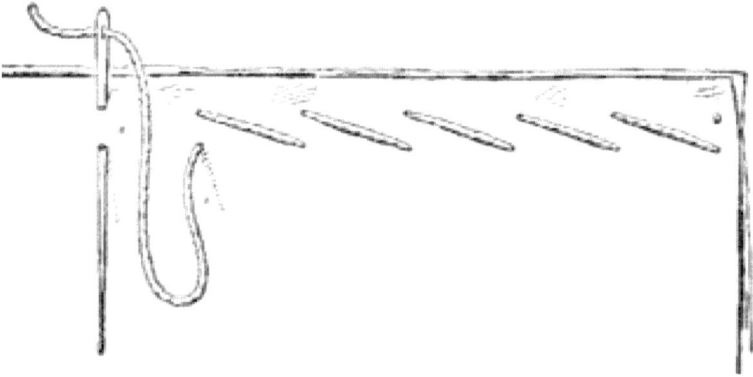

Diagonal basting

NOTE: It is found that to use long and short stitches when tacking holds the fabric more firmly. A knot is used to fasten on a tacking thread and a double stitch to fasten off.

2.4.2.2 Permanent Stitches
These are stitches essential in the construction of a garment, each with a particular purpose for which it should be used.

The quality of the completed garment depends on the high standard of hand finishing. However, for speed, certain of these processes can be worked by machine with the use of the appropriate attachments.

Machine Stitching (plain)
Used for speed to replace back stitching.

Hand Stitching
All hand sewing should be neat and firm with even stitches. The size of the stitch has to be adapted to the thickness of the fabric. Before starting, experiment with a few stitches to find the most suitable size for the fabric.

Fasten on thread with 2 or 3 small back stitches on sewing line. Fasten off with 2or 3 small stitches worked into the back of the completed stitching.

1. Running Stitch
Used for gathering, tucking, hand-sewing of fine fabric. Stitches and spaces between should be of equal length. Worked from right to Left

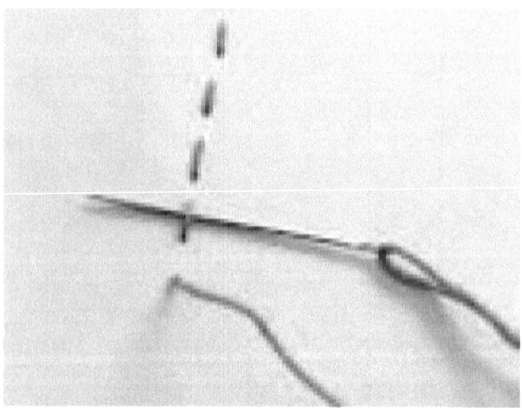

Running stitch

2. Back Stitch
Back stitch is a hand stitch that resembles a machine stitch on the right side and chain stich on the wrong side. Its strength and durability is equivalent to that of a machine stich. It has been in use before sewing machine invention. The stitch can be used on a crossway strip without difficulties. The stitch is used on pajamas,

Method: It's worked from right to left. A small crease is made on the edge of the two layers of fabric that will be stitched together. A small boarder is left at the top and bottom and small stitches made close together.

Back stitch step 1

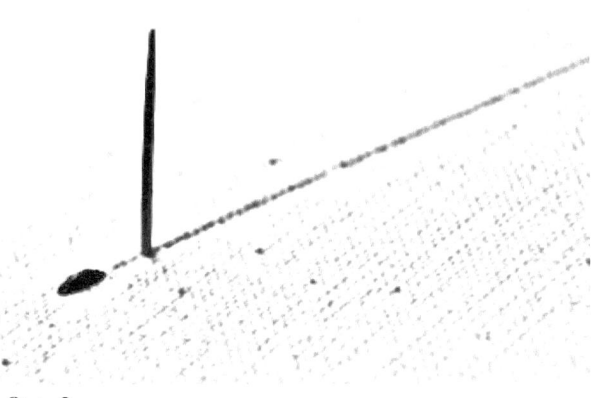

Step 2

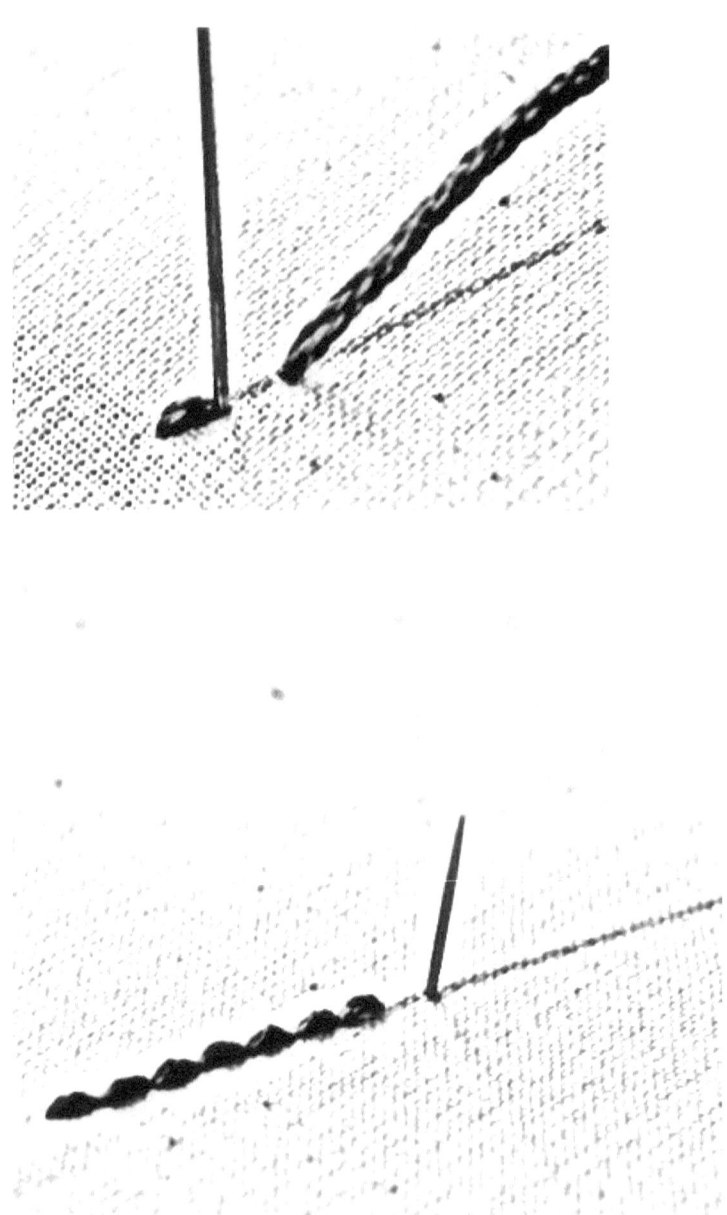

3. Oversewing worked from Right to Left

Used for joining two folded or selvedge edges, neatening of two edges of fabrics, details of construction and strengthening, joining and application of lace.

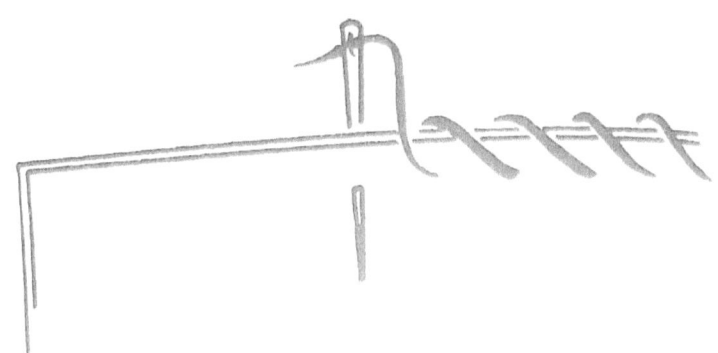

4. Overcasting worked from Left to right

Used to neaten raw edge of single layer of fabric.

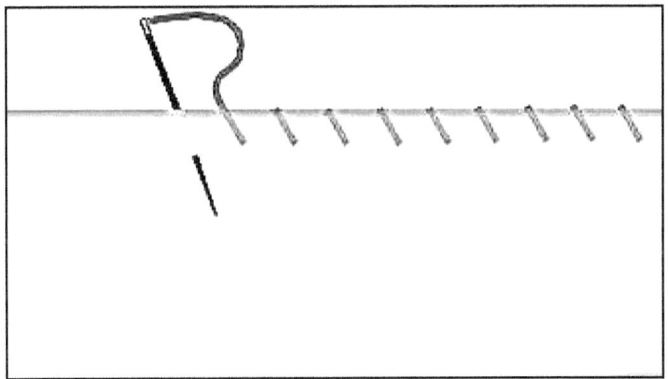

Hemming

Hemming is used on close to all garment hems. Heighten the beauty of a garment thus is one of the most important stitch. This is not visible on the right side of the garment or very small stitches may be seen.

Method: hemming is stitched by a single thread and a single thread is taken from the turned in surface of the fabric. The needle is passed through the single thread and through the surface of the fabric to give a decorative edge. Hemming can be done on sleeves, neck, skirts etc.

Slip Hemming
Slip hemming is similar to simple hemming but the stitches are made at small intervals from each other. It's normally used on slippery fabrics made from smooth fibers like silk and nylon.

Method: As the name suggests, the stitch resembles hemming but slanting (slipping) position. Used to neaten cuffs and necklines.

Blind hem

As the name indicates this stitch is nearly not visible to the naked eye. It has to be stitched with a lot of care to give a great finish. It is used particularly on men's clothes.

 Method: The turned in part is stitched very close to the main garment so as to pick a single strand of thread only at a time giving it nearly an inconspicuous feel.

Catch Stitch
Catch stitch is used to holding single edges such as interfacings. The stitch is not visible on the right side pick up single thread of fabric only when working on single fabric

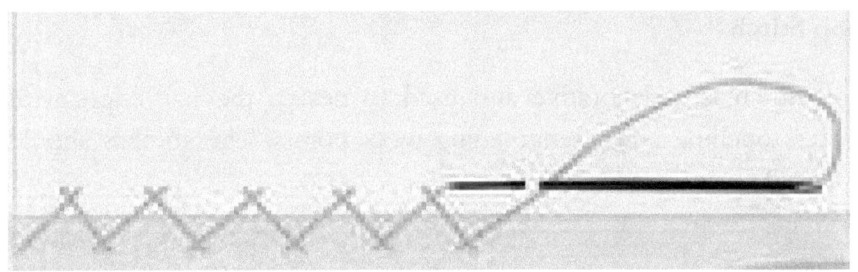

Buttonhole Stitch

Buttonhole stitch is very significant in the dressmaking trade, as it is used on nearly all kinds of garments –ladies, men's and children. A buttonhole is always cut on the top overlap. The buttonhole stitch is meant to neaten the buttonhole. It is sewn on two or more thickness of fabric. A buttonhole has a little curve on one side called a fan and an edge on the other called a bar.

Method: First decide on the intervals between each buttonhole. Maintain the diameter of the button, use the pointed end of a scissor to cut the buttonholes in the fabric. To ascertain that do not fray out neaten the raw edge with a temporary stitch. Buttonhole is always cut on a straight grain line. Then use a single thread to neaten the raw edge with a buttonhole stitch maintaining slightly extra tension on the 'fan' side to make a form of a chain stitch. Then pressed once the buttonhole is fully neatened.

RS   L to R

Loop Stitch

Loop Stitch is a decorative and used to neaten the raw edges with stitches touching and strengthening weak points. The stitches should be spaced correctly to form a square.

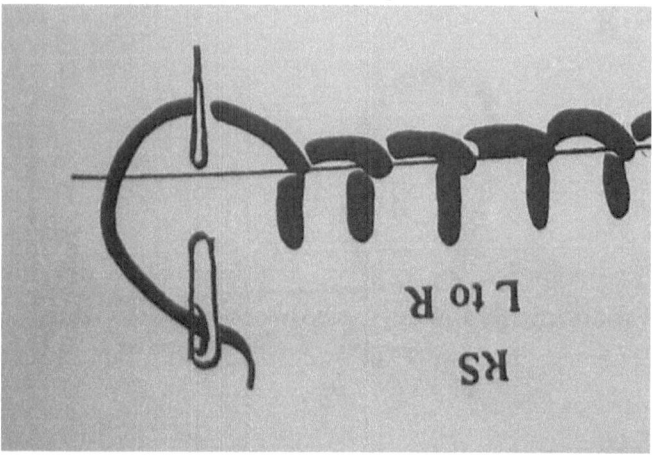

Herringbone Stitch

Herringbone Stitch decorative used to holding hem of thick fabric. It's worked across a single raw edge.
NOTE: Placing of the needle position.

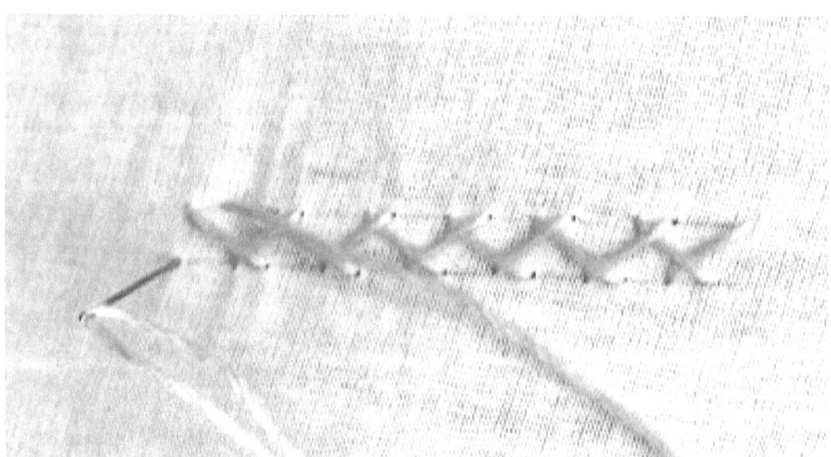

2.5 Revision Questions

1. Which section of the body is not measured to draft a skirt?
   A. Waist
   B. Hip
   C. Bust
   D. Skirt length
2. Which of the following body measurements are not taken horizontally?
   A. Waist length
   B. Waist bust
   C. Shoulder
   D. Hip
3. Why is it necessary to make adjustments to a paper pattern?
   A. To make a miniskirt
   B. To use less fabric
   C. To make measurements correspond and fit personal proportions
   D. Reduce the cost of the garment
4. Choose the correct method of reducing a skirt length of a paper pattern
   A. Make a pleat of width equal to half of the amount to be reduced, pin and trace the skirt
   B. Allocate more measurement for hem allowance
   C. Cut the extra length
   D. Avoid using the pattern
5. Which one of the following statements explains the correct meaning of one way fabric?
   A. Fabric that is printed on one side only
   B. Fabrics whose printed or woven images or patterns face one direction. It allows patterns to be cut while facing one direction.
   C. Fabrics whose printed or woven images or patterns face opposite direction
   D. Fabric that will stretch towards one direction

6. Which one of the following methods is not used to mark out patterns?
    A. Tailor marking
    B. Tailor's chalk
    C. Carbon paper
    D. Quilting
7. Which one of the following is true about temporary stitches?
    A. Temporary stitches are made on the right side of the fabric.
    B. Ate use to attaching two or more pieces of cloth together
    C. Stitches made to mark patterns, or hold layers of fabric together.
    D. They are considered to be very strong
8. Which one of the following is the odd one out.
    A. Oversewing
    B. Hemming stitch
    C. Basting stitch
    D. Loop stitch
9. Why do we add allowance while taking measurements?
    A. Allowance is added for ease of measurement
    B. To allow for cutting
    C. To ensure that the garment fits
    D. For shaping armholes and necklines
10. Which one of the is not an aspects of design to put into consideration when designing a garment
    A. Purpose of your garment/ dress
    B. Style
    C. Colour
11. Which one of the following is not a consideration when laying patterns on checked or striped fabrics
    A. Ensure that fabric is folded correctly so that all lines are exactly in position over one another
    B. Place the pieces to be laid on the fold line and the other pieces accordingly

C. Place the balance marks centres and hem lines of the sections to be joined in identical positions

D. Ignore the grain line to maximize on the fabric.

12. Which one of the following guides is considered when making hand stitches

    A. All hand stitches should be neat, firm and with even stitches

    B. All hand stitches should be made with a contrasting thread to make them decorative.

    C. The size of the stitch to be adopted to the thickness of the fabric

13. Which one of the following type of stitches is not hand embroidery stitch?

    A. Herringbone stitch
    B. Back stitch
    C. Overcasting
    D. Loop stitch

Supplementary Content

1. Measurement conversions

Yards to Metres		Inches to Centimetres			
Yards	Meters	Inches	Centimeters	Inches	Centimeters
¼ yd	0.229 m	1"	2.54 cm	21"	53.34 cm
½ yd	0.457 m	2"	5.08 cm	22"	55.88 cm
¾ yd	0.686 m	3"	7.62 cm	23"	58.42 cm
1 yd	0.914 m	4"	10.16 cm	24"	60.96 cm
1 ¼ yds	1.143 m	5"	12.70 cm	25"	63.5 cm
1 ¾ yds	1.60 m	6"	15.24 cm	26"	66.04 cm
2 yds	1.82 m	7"	17.78 cm	27"	68.58 cm
2 ¼ yds	2.058 m	8"	20.32 cm	28"	71.12 cm
2 ½ yds	2.286 m	9"	22.86 cm	29"	73.66 cm
2 ¾ yds	2.515 m	10"	25.4 cm	30"	76.2 cm
3 yds	2.743 m	11"	27.94 cm	31"	78.74 cm
3 ¼ yds	2.972 m	12"	30.48 cm	32"	81.28 cm
3 ½ yds	3.429 m	13"	33.02 cm	33"	83.82 cm
3 ¾ yds	3.429 m	14"	35.56 cm	34"	86.36 cm
4 yds	3.558 m	15"	38.1 cm	35"	88.9 cm
4 ¼ yds	3.887 m	16"	40.64 cm	36"	91.44 cm
4 ½ yds	4.115 m	17"	43.18 cm	37"	93.98 cm
4 ¾ yds	4.344 m	18"	45.72 cm	38"	96.52 cm
5 yds	4.572 m	19"	48.26 cm	39"	99.06 cm
-	-	20"	50.8 cm	40"	101.6 cm

2. Table of measures

2.1 Table of measures for men

Breast	30	32	34	36	38	40	42	44	46	48	50
Waist	28	29	30	32	34	37	39	42	45	48	52
Seat	33	35	36	38	40	42	44	46	48	50	52
Waist-length	15 ½	16	16 ¼	16 ½	17	17 ¼	17 ½	18	18 ¼	18 ½	19 ¾
Back-width	6 ¼	6 ½	6 ¾	7	7 ½	8	8 ¼	8 ½	9	9 ¼	9 ½
Sleeve - length (with back-width)	30 ½	31	31 ½	32	32 ½	33	33 ¼	34	34 ¼	34 ½	34 ¾
Depth of Scye	7 ¾	8 ¼	8 ½	9	9 ½	9 ¾	10	10 ¼	10 ½	11	11 ¼
Across chest	6 ½	7	7 ½	8	8 ½	8 ¾	9 ¼	9 ½	10	10 ¼	10 ½
Front shoulder	11 ¾	12	12 ½	13	13 ¾	14	14 ¼	15	15 ½	16	16 ½
Over-shoulder	16	16 ½	17	17 ¾	18 ¼	18 ¾	19 ½	20 ¼	21	22	23
Neck	13 ½	14	14 ½	15	15 ½	16	16 ½	17	17 ¼	18	18 ½
Trousers-length	39	39 ½	40	41	42	43	44	45	46	46 ¼	46 ½

2.2 Table of measures for women

Chest	26	28	30	32	34	36	38	40	42	44
Waist	24	24	24	24 ½	25	26	27	28 ½	30	31 ½
Seat	26	29	32	36	38	40	42	44	46	48
Waist-length	11	12	12	14	15	15	15 ¼	15 ½	15 ½	15 ¾
Back-width	5	5 ½	6	6 ¼	6 ½	6 ¾	7	7 ¼	7 ¼	7 ½
Across chest	5 ½	6	6 ½	7	7 ½	8	8 ¾	8 ¾	9 ¼	9 ½
Depth of Scye	6	6 ½	6 ¾	7	7 ¼	7 ½	7 ¾	8	8 ¼	8 ½

2.3 Table of measures for boys

Age	4	5	6	7	8	9	10	11	12	13	14
Breast	23	23 ½	24	25	26	27	28	29	30	31	32
Waist	23 ½	23 ½	24	25	26	27	27	27 ½	28	29	30
Waist length	10	10 ¼	10 ½	10 ¾	11 ½	12 ¼	12 ¾	13 ¼	13 ½	13 ¾	14 ½
Back-width	5	5 ¼	5 ¼	5 ½	5 ¾	6	6	6 ¼	6 ½	6 ¾	6 ¾
Sleeve length	18 ½	19 ½	20 ½	22	23 ½	24 ¾	26	27	27 ½	28 ½	20 ¼

2.4 Table of measures of yardage

Garment	Breast	Full length	Sleeve	Cloth-width	Yards
Full shirt	36	32	23 ½	30	3
Full shirt	24	23	16	27	1 ¾
Half shirt	36	32	11	30	2 ½
Fashionable blouse	36	16 ½	8	36	1
Girl's Polka	24	15	6 ½	27	1
S.B. Lounge jacket	36	27 ½	24 ½	27	3 ½
Waistcoat	36	25	…	27	¾
Jacket & vest	36	27 ½	24 ½	27	4
Overcoat	36	42	24 ½	56	2 ¼
	Seat		Bottom		
Trousers	38	42	24	27	2 ¾
Riding breeches	38	34 ½	10	27	2 ¼
Jodhpurs	38	42	12 ½	27	2 ¾

Chapter 3: Seams

3.1 Introduction

Seams are requirements for all garments and it is important that one chooses the best type of seam. The type of seem chosen should be used throughout the garment. Good preparation, stitching and neatening of the seams plays a big role in ensuring that the seams are made to their required standards.

The Seams will either show on the right side (conspicuous) or not show on the right side (inconspicuous). Conspicuous seams can occasionally be as decorative detail.

Choice of seam depends upon the following points:

   a. Weight of the fabric being used
   b. Position and purpose of seam
   c. Type and purpose of garment
   d. The style line of a seam

General guides to consider when making all seams:

1. Match balance marks and fitting line carefully and pin work into position correctly.
2. Tack the layers of fabric firmly along fitting line and remove pins to prevent the layers of fabric from out of position when stitching.
3. Machine stitch exactly against the fitting or tacked lines so that these are not caught into the permanent stitching and can be removed easily.
4. Press each stitched line first and then press turnings open or to one side as necessary.
5. It is important to ensure that a sewing machine has elasticity when seaming knitted or stretch fabrics.

3.2 Types of seams

3.2.1 Plain Seam: inconspicuous

The easiest and flattest method of seaming used generally but in particular for fabrics that do not fray and on thick fabrics since the seam will not appear bulky. This seam is not self-neatened thus requires neatening to prevent the raw edges from fraying. The most suitable method of neatening should be chosen according to fabric and position of seam

1. Place right sides of the pieces to be joined together while matching balance marks and fitting lines carefully. Pin and tack in position

2. Stitch along the fitting lines, remove tacks and press stitching.
3. Pressing turnings open and neaten the raw edges.

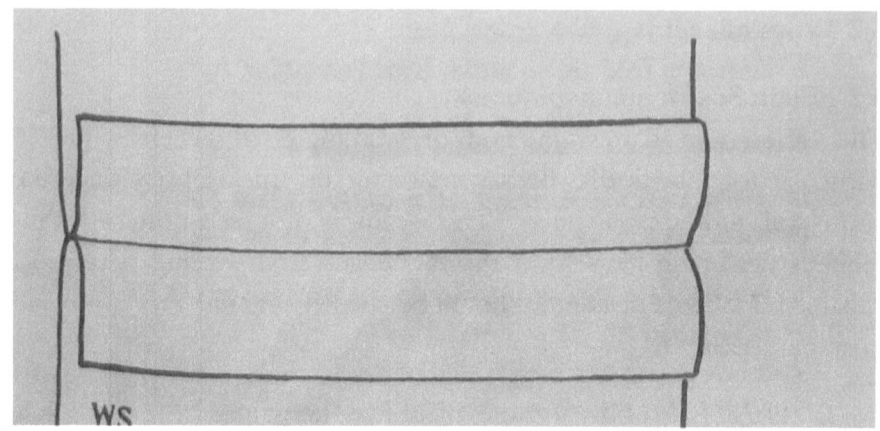

Methods of Neatening of Plain Seam
Flat Edge Neatening

1. Trim seam turnings equally on both sides.
2. Neaten with the most appropriate method
(a) Hand or machine stitched overcasting.
(c) Loop stitch.

Edge Stitching:

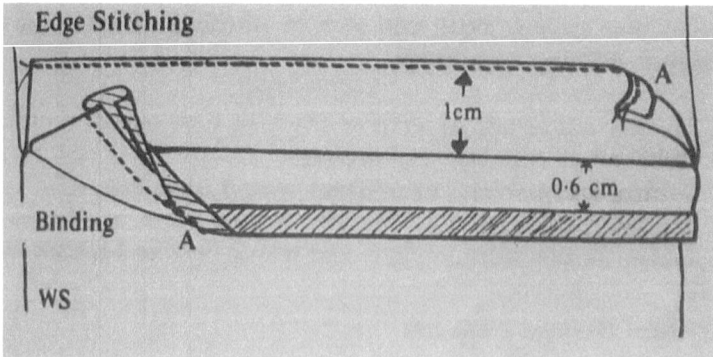

With right side of seam turning uppermost, fold the raw edge under to a line 1 cm from stitching line. Stitch 0.15 cm from folded edge and press.

Binding:

This method is used thick fabrics that fray easily. Thin silk fabric or fine binding can be used. A contrasting colour of binding may be used to make it decorative.

1. Mark and cut the crossway strip
2. Place right side of strip to right side of turning, pin and tack into position so that the stitch line is 0.6 cm from seam line. Stitch, remove tacks and press.
3. Press binding strip upwards and fold strop over the raw edge to the wrong side.
4. Turn in raw edge of strip and place fold against stitch line.
5. Hem the fold on to stitch line.

Neatening of a Curved Seam

The turnings of a curved plain seam snipped in order to make flat when pressed open,

1. Snip cross turning to 0.3 cm of stitched line.
2. Neaten the raw edge with a loop stitch.
3. Press turning open.

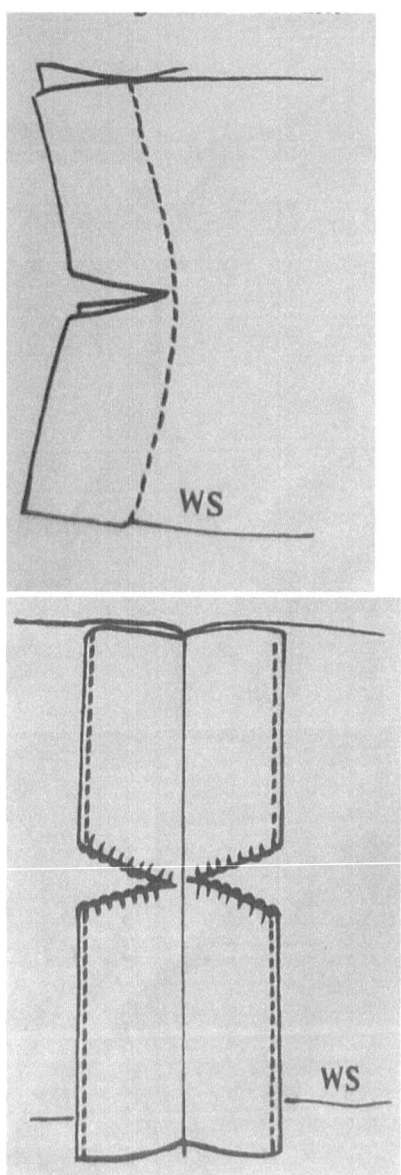

Methods of Neatening Waist and Armhole Seams

In many cases these seams are joined with a plain seam. Turnings in these areas are neatened together and not pressed open. Turnings of other seams and darts are correctly placed: these should have been completed and neatened before waist and armhole seams are joined.

The following are suitable for neatening waist or armhole seams.

Hand-stitched Neatening

Hand stitching gives the smoothest finish to seam that may be very bulky in places. To give the neatening stitch extra strength, a further line of machine stitching is necessary.

1. On both layers of turnings work a second line of machine stitching 1 cm outside the fitting line.
2. make overcasting over the raw edges

3.2.2 French Seam: inconspicuous

This seam is mainly used for lingerie, fine blouses and children's wear. If possible it should always be used on sheer fabrics such as chiffon when used singly, i.e. not mounted on to a lining fabric. It can only be used on fine fabrics as otherwise it will be too bulky.

A french seam is strong and self-neatening as all raw edges are enclosed; therefore, it launders well. Finished width of seam must not be more than 0.6 cm or less on fine or sheer fabrics.

1. Carefully place wrong sides together, matching balance marks and fitting lines.
2. Machine stitch 0.6 cm outside the fitting line. less on very fine fabrics.

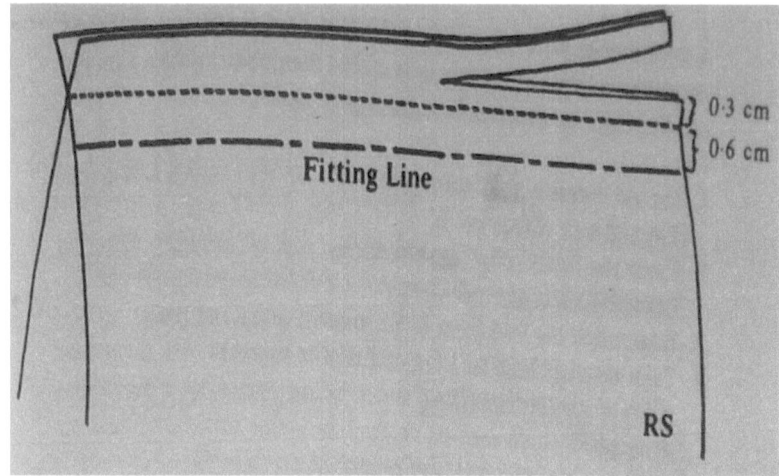

3. Press turnings open at this stage to make the following stages easier and produces a successful finish.
4. Turn the work through to wrong side fold back on the stitched line and tack fitting lines together again. Stitch on fitting line enclosing raw edges.

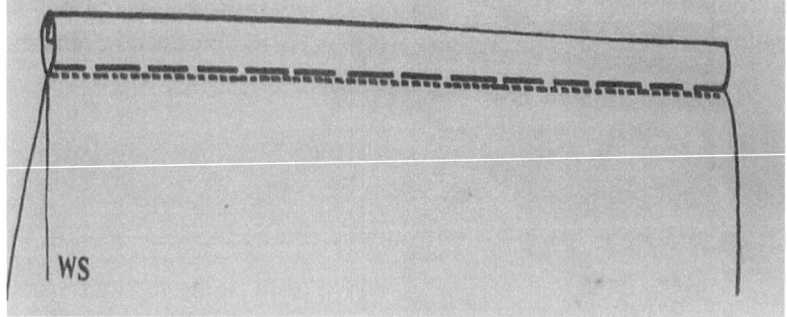

5. Remove all tacks, press stitching and then press seam over towards the back of the garment.

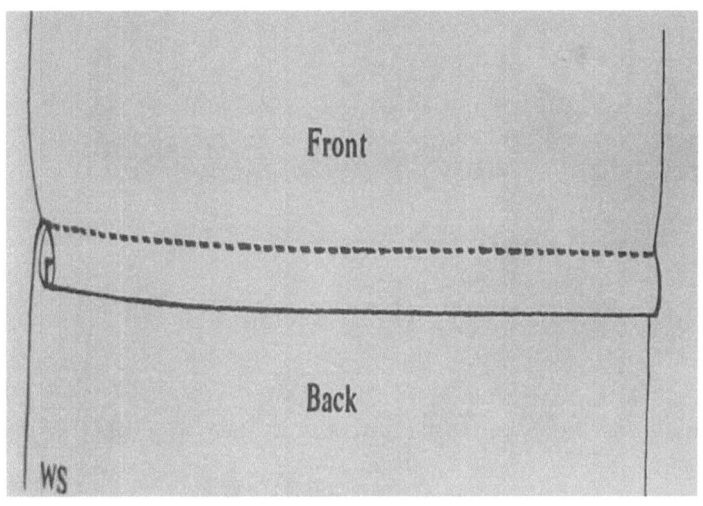

3.2.3 Overlaid Seam: conspicuous

A strong seam used for joining difficult shaped piece, e.g. yokes and shaped panel lines. A stitching is visible it emphasizes style lines and can therefore be planned as a decorative detail.

1. On the overlay, fold seam allowance to W.S. along fitting line. Pin and tuck as shown on the diagram below.

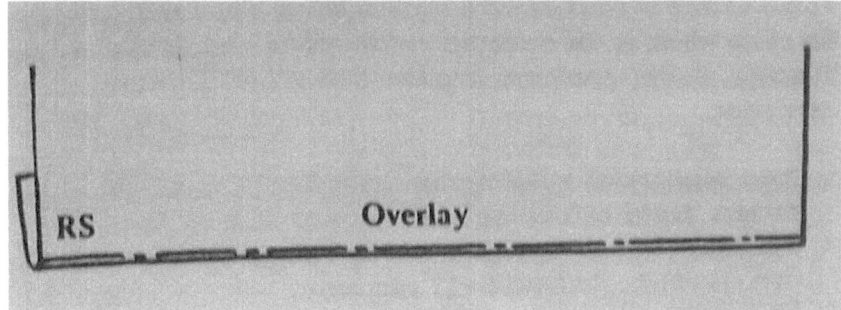

2. Place the folded edge against fitting line of underlay, matching balance marks. Pin and tack.
3. Edge stitch the fold from R.S.; remove tacks and press.
4. Trim turnings level to 1 cm and neaten together with either loop stitch or oversewing as illustrated below.

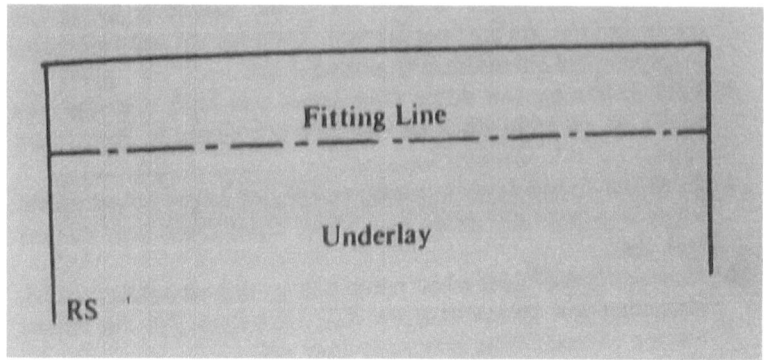

When neatening, a second line of machine stitch may be worked as in the method of finishing waist and armhole seams.

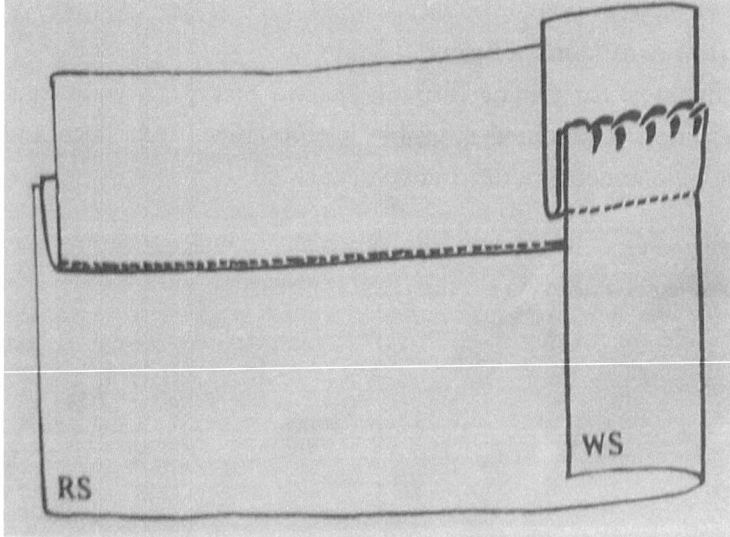

On children's wear, blouses and lingerie, a decorative stitch is sometimes used instead of edge stitching. Suitable stitches would be stem, chain or interlaced stitch.

3.2.4 Machine and Fell: conspicuous

This is known as run and fell if worked entirely by hand; it is a strong flat seam which is self-neatened and launders well. Is used on shirts, pyjamas, shorts, trousers, overalls and other garments receiving harder wear.

1. Place together with R.S. facing the pieces to be joined, matching balances marks and fitting line. If worked by hand use a close running stitch. Remove tacks and press.
2. Trim seam allowance of back or lower section of garment to 0.6 cm (a little less for finer fabrics). Trim second seam allowance evenly to 1.25 cm width as shown on the diagram below.

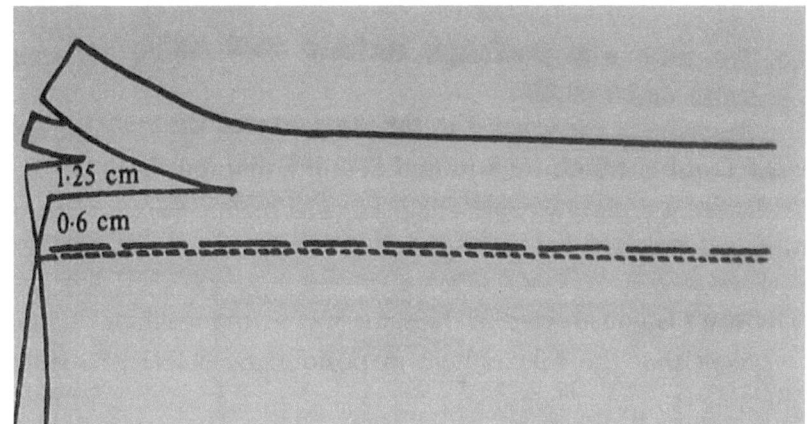

3. Fold projecting raw edges over lesser one with fold line even width 0.6 cm from stitching and tack Press folded edge. As illustrated below.

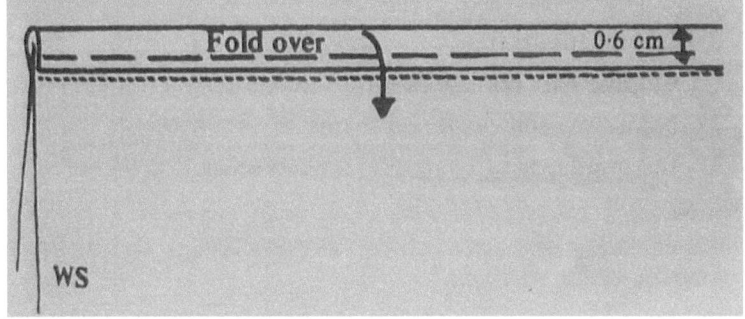

4. Open out fabric so that joined pieces are flat and press folded edges into position: press iron away from stitch line. Pin and tack flat.
5. Hem neatly the folded edge: remove tacks and press. Hemming will be invisible on R.S.

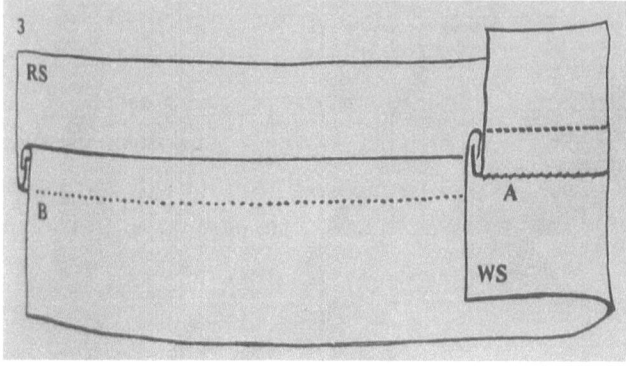

3.2.5 Double Machine Stitched Seam: conspicuous

A flat seam with raw edges enclosed used mainly for trousers, pyjamas and overalls.

The seam is constructed in the same way as the machine and fell seam – except that the fold is held in position by a line of top machine stitching instead of hemming.

When joining different sections of the garment note the correct position of the overlap.

3.3 Revision Questions
1. Outline the consideration to make when choosing a seam
2. Name the self-neatened seams
3. List methods of neatening a plain seam.

Chapter 4: Disposal of Fullness

Disposal of fullness is the arrangement of extra material in a garment. Fullness is a significant characteristic of fashion on a garment, it facilitate movement in a close-fitted garment. Fashions changes based on the ways of controlling fullness adapted to enhance the style.

4.1 Darts

A dart is a stitched fold of material which is tapered to a narrow point. They are used to make a garment fit smoothly, give room for movement where necessary and dispose of fullness where desired at the waistline. Darts are often arranged to form part of the style lines of the garment.

4.1.1 Single Pointed Dart

Matching fitting line carefully fold fabric Right Sides together. Pin and tack.

Machine stitch starting from the wide end on fitting line until the fold is reached, and then continue for three more stitches along the fold.

(Careful pressing, will ensure a smooth finish on R.S.)

Remove tacks and press waist darts towards the Centre Front or Centre Back line. Press under-arm darts towards waistline, shoulder darts towards the neck and elbow darts towards the wrist.

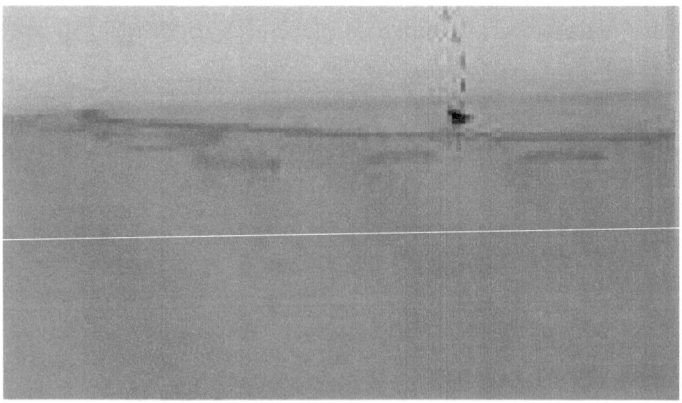

4.1.2 Double Pointed Dart

Double pointed darts are used on dresses and coats which have no waistline seam.

Matching fitting line carefully fold fabric Right Sides together and tack.

Machine stitch three stitches on the top point of dart along the fold. Follow the fitting line to the waistline and down to the other point and along fold for three more stitches.

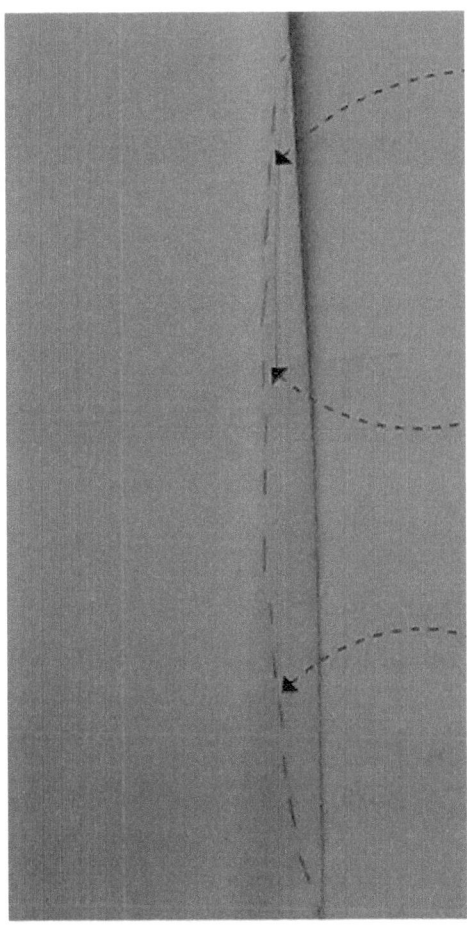

Remove tacks and press stitching.

Snip the dart 0.3 cm at the waistline to prevent dragging on the Right Side of the stitching and neaten the raw edges.

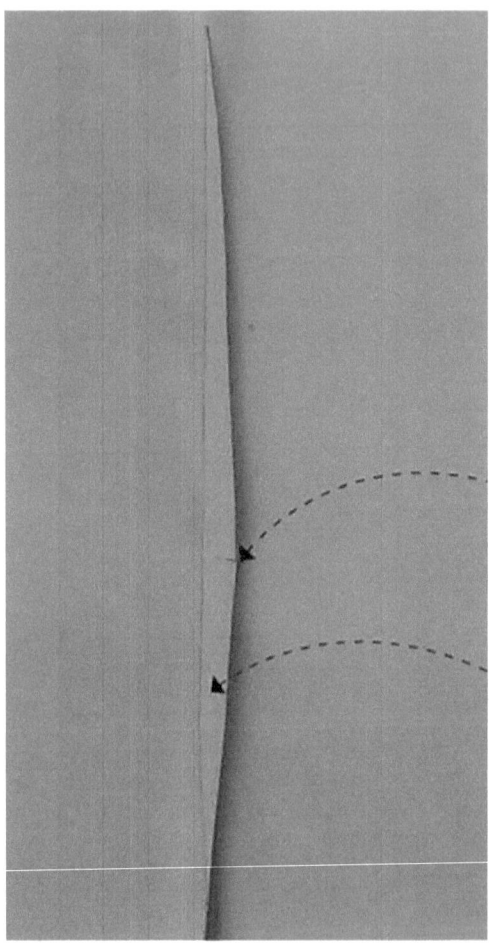

4.1.3 Dart Tack

These are darts stitched half way to leave open-ends similar to tucks. These kinds of darts are good for styling they are used for waist darts on blouses and shoulder darts on coat and jacket linings.

1. Matching fitting line carefully fold fabric together and tack.
2. Starting from the wide end, machine stitch on the fitting line to end of dart tuck, reverse machine stitching for 1.25 cm to finish and strengthen the stitching.
3. Remove tacks, press stitching and press to correct position.

4.2 Pleats

A pleat is an unstitched fold of material held in place along a conjoining seamline. It is a fold in the fabric that releases fullness. Pleats are used to create room for movement or as a feature of design.

There are three main forms of pleats:

1. *Knife pleats:* are equal with pleats folded towards the same direction.
2. *Box pleats:* are two knife pleats folded facing opposite each other.
3. *Inverted pleats:* are two knife pleats folded facing each other.

The amount of fabric required is three times the finished width of pleat. Therefore, for a skirt with knife pleats all round three times the finished hip measurement plus turnings must be allowed for.

The basic methods of preparing these are shown in a simplified form. It will be found when following a commercial pattern that pleats are either (a) slightly flared 1.25 cm wider at the hem and a little narrower at the waist which gives and excellent hand and switch to the pleats, or (b) that they are overlapped or widened between hip and waistline in order to decrease the hip measurement to that of the waist.

For all pressed and stitched pleats, the edge of the fold should be thread-marked and accurately pressed back from waist to hem *before* assembling the pleats.

The arrow indicates the direction of fold to be made.

When the waistline (or setting-in of pleated panels has been prepared, try on the garment to check that the pleats hang correctly; if they appear to drag, the deep inside fold may have to be lifted slightly to correct this before the final stitching. Firm waist finishes are required for pleated skirts of heavier fabrics.

After pleating has been finished, baste the pleats into position until the garment is completed.

4.2.1 Knife Pleats

These pleats all fold in the same direction.

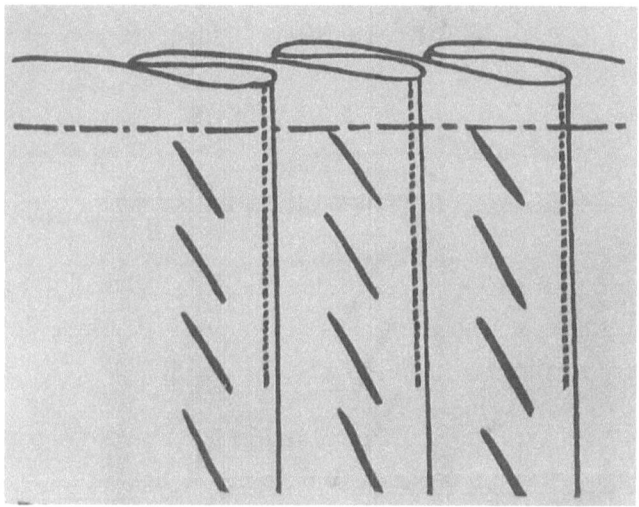

4.2.2 Box Pleats

These are made by two knife pleats folded facing opposite each other (which forms an inverted pleat on W.S.)

Mark position of pleats:

Fold and press carefully each pleat edge A' to line A. Pin and diagonally baste into position.

Fold edge B to line B': Pin diagonally baste.

Stitch on fold from required length to waist edge. Fasten ends off securely.

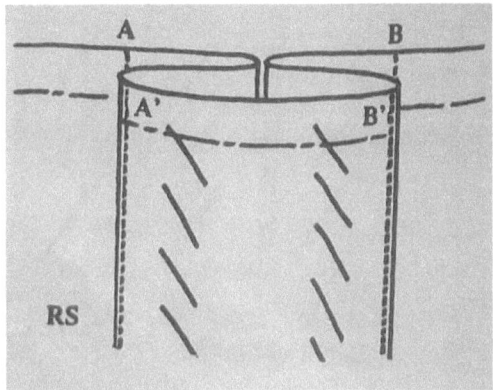

These are made by two knife pleats folded facing each other (which forms a box pleat on the W.S.).

Mark position of pleats:

A – B = twice depth of pleat.

B – A' = twice depth of pleat.

A' – C = twice depth of pleat = distance between inverted pleats on R.S.

Fold and press accurately each pleat edge A, A', C, C'.

Fold edge A to line B. Pin and diagonally baste into position.

From top of pleat stitch folded edge to length required, turn fabric under machine and stitch upwards on second folded edge.

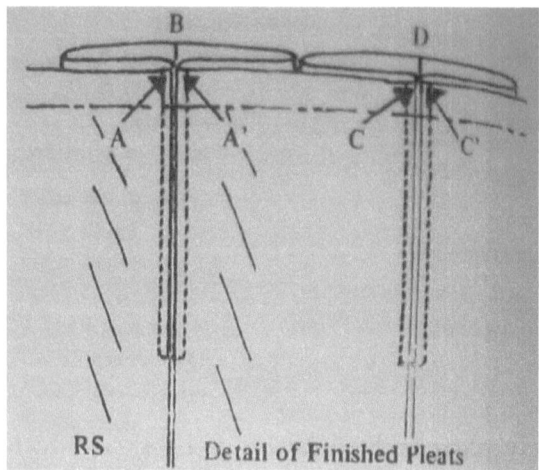

Detail of Finished Pleats

4.2.4 Kick Pleats

This pleat, which has no stitching visible on R.S., allows for movement in a straight skirt and is inserted at the base of centre back seam, side or panel seams. Patterns usually allow additional seam allowance for this pleat. If the fabric pulls easily and the pleat is liable to strain, an arrowhead should be worked in for reinforcement of seam ending.

1. Complete the machine stitching for the plain seam as far as the pattern mark. Do not remove the tacking which holds the unstitched section in place.
2. Place the underlay of the pleat into position. Pin and tack on to the seam allowance.

Machine stitch underlay to each side of seam allowance and across the top of the pleat

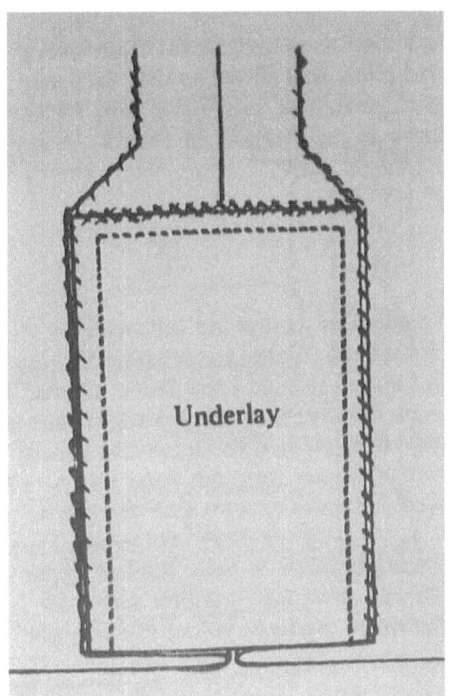

Neaten the raw edges by the same method and as a continuation of the seam neatening. When edge stitch neatening has been used generally in the garment, neaten raw edges of pleat with over-sewing. Remove tacks and press. If necessary, complete with an arrowhead on R.S.

Arrowheads are used to strengthen the seam ending at the top of a kick or inverted pleat. It is placed so that the lower straight edge is level with top of pleat. Use buttonhole twist for stitching. Method of work is shown in the diagram, first by marking the outline of the triangle shape.

4.3 Tucks

Tucks are small pleats that evenly spaced and stitched to the required depth. They are decorative methods of controlling fullness of a garment. Tucks are used on children's to hold extra fabric allowed for growth, particularly around the skirt hem. Often completely tucked fabric is used to form decorative panels.

Width of the tucks change from pin tucks 0.15 to 1.25 cm and are machine stitched, hand run or sewn with decorative stitch

Allowance of extra fabric is twice finished width of each tuck. Measure, stitch and press into position each tuck before starting the next. Complete all the tucks before marking out the pattern or fitting lines.

4.3.1 Markers

A cardboard maker with cut notches is used to show the required width of completed tuck and the edge of the next tuck. Ready-made makers can be bought or easily made at home.

4.3.2 Wide Tucks

1. Prepare markers as required. Start work from the side to which tucks are to be pressed.

Fold the fabric along the thread R.S. out for the edge of the first tuck. Tack width of tuck.

Stitch from top on R.S. of tuck. Remove tacks and press stitching.
Open out fabric and press tuck flat into position.

2. With fold towards marker, as at point A, place marker correctly and fold back fabric for edge of second tuck and tack width of tuck.

Stitch from top side of tuck, remove tacks and press.
Open out fabric and press tuck flat into position.
Repeat.

4.3.3 Pin Tucks

These are tiny edge stitched tucks 0.15 cm wide, sewn either by hand running or machine stitching. Prepare marker as required and start work from the side to which tucks are to be pressed.

Fold fabric along the thread R.S. out for the edge of first tuck. Tack if necessary and edge stitch from top on R.S. of tuck.

Press stitching. Open out fabric, press tuck flat into position.

With fold towards marker as at A, place marker correctly and fold back fabric for second tuck. Edge stitch and press. Open out fabric and press tuck flat into position. Repeat.

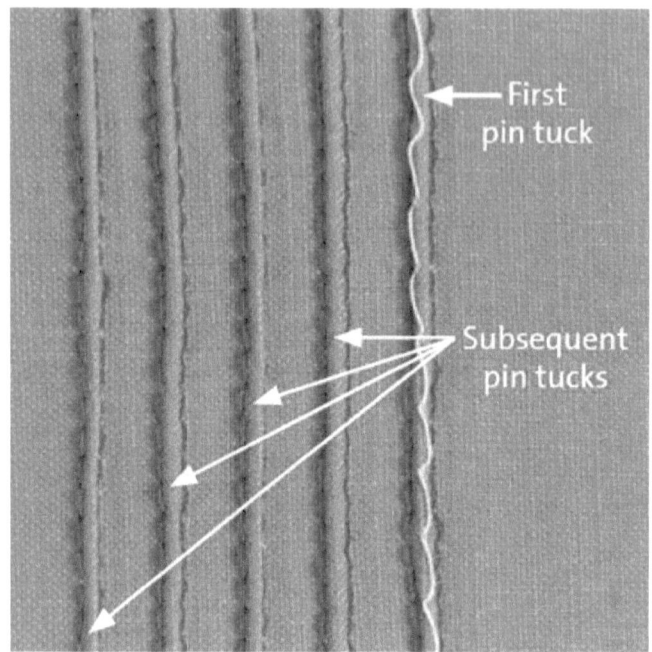

4.3.4 Decorative Tucks

Shell Tucks

Desirable for blouses and children's wear of light fabrics. Follow method given for wide tucks.

1. Prepare and tack width of tucks 0.6 cm or less

Work shell edging from top on R.S. of tuck. Open out fabric and lightly press into position.

Prepare second tuck and repeat process.

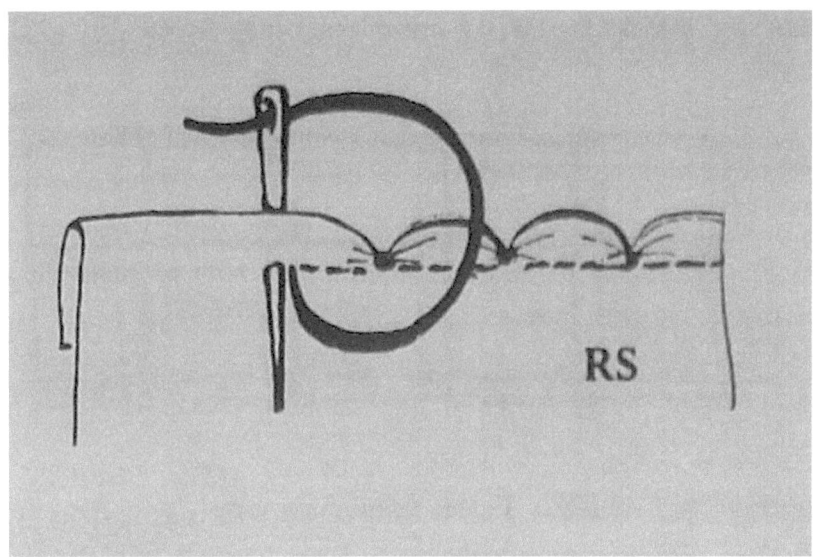

Hem-stitched Tucks

Desirable for lightweight fabrics woven loosely which will allows threads to be withdrawn easily.

1. Mark edge of the tucks
2. Draw out a thread along the width of tuck on each side of fold line. Draw more threads for width needed towards the fold line, so that two lines of drawn threads exactly opposite.
3. Tack into position. Work hemstitching from underside of tuck and press.
4. Open out fabric and press tuck flat into position. Measure distance for fold of second tuck as method given for wide tucks. Mark edge of fold with tack line and repeat process.

Wide tucks can also be stitched decoratively with stem, chain, whipped, running stitches or French knots.

Methods of tucking lightweight fabrics: suitable for large areas or whole garments such as over blouses and dresses: tack all fabrics before marking pattern, using machine stitching for speed.

Cross tucking: pin tucks. Tuck fabric in one direction first and then at right angles. Effective when styles to be cut on the cross.

Graduated tucking: wide tucks. Increase depth of tucks towards hem.

4.4 Gathering

Gathering is controlling fullness with even fine unstitched tucks. This is applicable on lightweight fabrics at waist, wrist, yoke and sleeve head. Gathered sections are set in either an overlaid seam or a plain seam which can have a decorative finish

Gathering is accomplished with the aid of two rows of gathering threads worked 0.5 cm below and above the fitting line

4.4.1 Gathering by Hand

Fasten on the thread firmly and make two rows of the running stitches. The stitches of the second row must be placed directly in line with those of the first row. For ease in drawing up, work the second row in the opposite direction so that the thread can be drawn from each end.

4.4.2 Gathering by Machine

The longest stitch with a loose tension must be used. Place stitches of second row in line with those of the first. Draw up one of the machine threads only for each row, otherwise the fabric will not move.

Drawing up: draw up both lines together, gently easing fabric along, until correct width is reached. Hold threads temporarily in place by twisting round vertical pin. Disperse gathers evenly.

4.5 Smocking

This is a neat method of controlling an area of fullness: as the method and tension of work is not rigid, it provides for elasticity and ease of movement.

The allowance of fabric required is usually four times the finished width of smocking. On very fine fabrics, a greater allowance may be needed and a small piece of fabric should be gathered and drawn up to check this. It is advisable to check all fabrics in this way as the amount required does vary according to the fabric and depth of pleat required. When drawn up the pleats should just touch but should *not* be tightly packed.

4.5.1 Preparation for Smocking

On check, striped or spotted fabrics, the pattern should be used as a guide for putting in the gathering thread.

On plain or freely patterned fabrics, a transfer of smocking dots can be used: these are available in yellow and blue and with different spacing according to depth of the section to be gathered.

Place the transfer on the W.S. with the spots in line with the thread of the fabric. With a moderate heat press the iron over the dots, peeling off the paper immediately after the iron has passed. On sheer fabric the transfer must be pinned into position and the gathering worked through the paper which is then torn away, as the transfer spots would be visible on the R.S. if ironed on.

The top line of gathering is placed 0.3 cm above fitting line. Gathering lines should be approximately 0.6 cm apart to hold the pleats firmly in position.

Use a sufficient length of strong cotton separately for each line of gathers and fasten on securely with a knot or a double stitch.

Working from R.S., start at the right-hand end of the top line and work row of gathering, picking up a small stitch behind each dot (for fine pleats) or passing needle in and out of dots. On striped or similar fabrics, plan the gathering to coincide with the pattern, i.e. to pick up the same coloured stripe. When drawn up, this would give a different coloured effect to the main fabric.

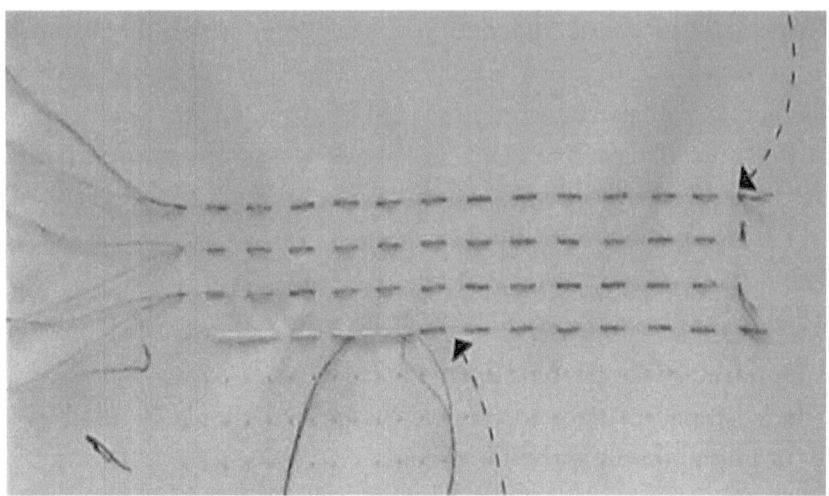

When all rows have been stitched, draw up in pairs, starting at the top. Until drawing up is complete, twist the gathered threads round vertical pins. When all rows have been drawn up evenly and the gathered area is the exact width required, tie off the loose ends.

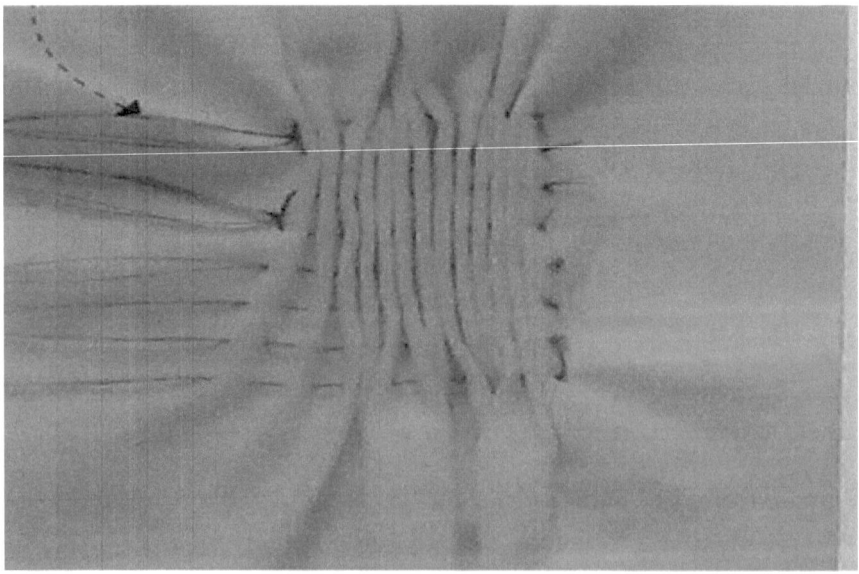

NOTE: Gathering threads must not be removed until smocking has been completed.

4.5.2 Smocking Stitches

Use a smooth firm embroidery thread such as cotton a Broder or a suitable number of stranded cotton threads or a twisted embroidery thread. Smocking is usually most effective when worked in one colour throughout. Fasten on all stitches on the W.S. by working two stitches over the second pleat, then push the needle through to R.S. between the first and second pleat. Work with an even, slightly loose tension so that smocked area does not contract, when the threads are withdrawn.

Stem Stitch

1. Work from left to right stitching through each pleat holding the thread above the needle
2. Do not pull thread too tight as this would reduce the final width of the smocking and prevent the elasticity
3. The chain stitch is achieved by turning round at the end and working back with the stitches close together

Cable Stitch

1. Work as for stem stitch but alternate the thread above and below the needle
2. The double cable is worked by stitching another row under the first, with the lower stitch of the first row touching the upper stitch of the second row

Wave Stitch

Work making sure that there is the same number of stitches (going down as coming up) on each side of the wave.

Working downwards the thread is above the needle and working upwards the thread.

4.6 Revision Questions

1. Define disposal of fullness

   Answer: Arrangement of the extra material in a garment

2. List the types of tucks

   Answer:
   a. Pin tucks
   b. Shell tucks

3. Which one of the following is not a method of disposal of fullness

   A. Tucks
   B. Slip hemming
   C. Gathering
   D. Smocking

4. Select a method of control of fullness that can be applied on a man's

   A. Gathering
   B. Smocking
   C. Tucks
   D. Darts

5. Which one of the following are not types of darts?

   A. Single pointed darts
   B. Double pointed darts
   C. Accordion dart

6. Which one of the following is not a type of pleat?

   A. Knife pleat
   B. Double pointed pleat
   C. Box pleat
   D. Inverted pleat

Chapter 5: Crossway Strips, and Facings and Interfacings

5.1 Crossway Strip
Introduction

These are strips of fabric cut across the grain at an angle of 45° to the warp and weft threads of a fabric. Making the warp and weft threads interweave at right angles (90°) to each other on a woven fabric. When the fabric is pulled diagonally from the straight edge, the woven threads move out of place, allowing it to stretch. The narrow strip of fabric cut diagonally at 45° to the woven threads will stretch when pulled; this will allow it to be manipulated around curved edges in order to lie smooth and flat.

These crossway strips are used as binding or facing for:

(a) Neatening Seams or hem edges
(b) Finishing Curved raw edges, armholes and necklines
(c) To give a decorative finish
(d) To attach collars and cuffs

A strip or panel that is cut at any angle other than 45° to the straight thread is known as 'on the bias'. A strip on the bias will not manipulate as well as a crossway because the pull of the thread is uneven.

Bias binding: this is a commercial name for a folded crossway strip wound on to a card which may be bought in any harberdashery department. It is available in various qualities of cotton and rayon: This prepared binding is particularly useful for

(a) Decorative finishes and

(b) When crossway strips cannot be cut from the main fabric of the garment.

5.2 Cutting and joining crossway strips

5.2.1 Cutting Crossway Strips
1. Straighten two sides of the fabric to be used by cutting to a thread along both warp and weft
2. Fold the straight warp edge across the fabric to form a right angle (90°) and parallel to the weft threads.
3. Cut through the fold formed from A to B, putting the scissor blade firmly into the fold
4. Measure in from the cut crossway edge the required width of the strip and mark with pins: a measure card is used for this.
5. Cut along marked line with care as crossway moves easily out of position.

5.2.2 Joining a Crossway Strip
A crossway strip should always be joined on a straight grain to make all seams fall in the same direction.

1. Cut all strip ends on the straight grain and in the same direction, i.e. all warp or all weft, forming a parallelogram.
2. Place strips in line R.S. up with straight cut ends parallel.
3. Turn strip A over on to B so that R.S. facing with the straight ends together. Allow a point of fabric to project 0.6 cm each side to give a seam allowance of 0.6 cm. accurately positioned a right angled 90° will be found at points X. Pin and tack.
4. Stitch, join with 0.6 cm turnings and fasten off ends securely. Remove tacks. Press seam open and trim off points. If a striped fabric is used, plan to match stripes exactly as shown in the diagrams.

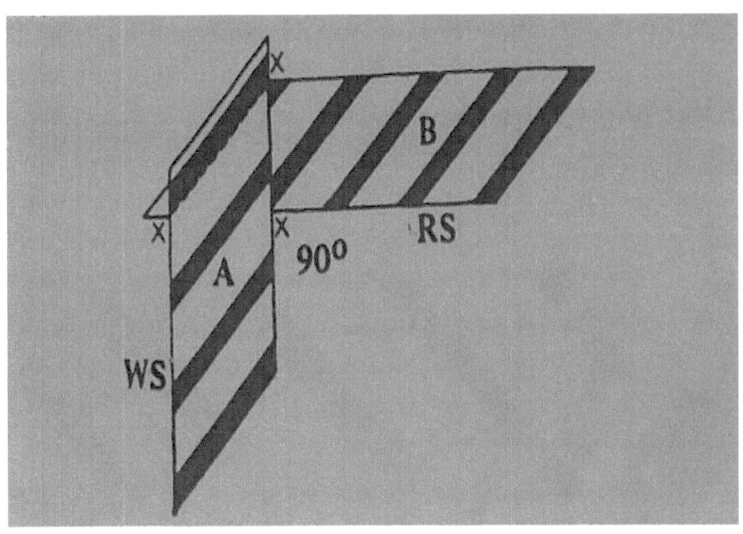

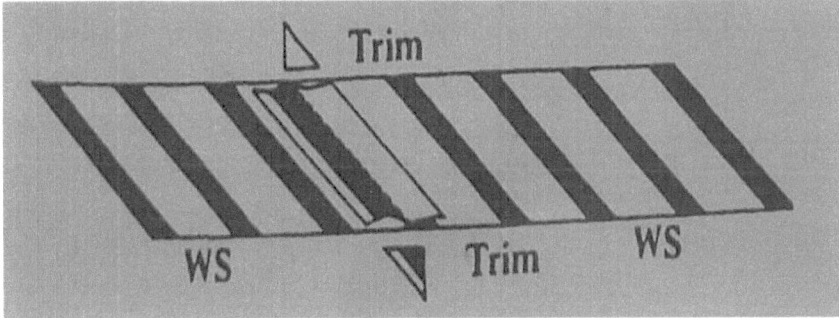

5.2.3 Crossway Strips used for Binding

Binding is a decorative method of finishing raw edge and neatening seams. They are also used on raw edges on the inside of garments, especially those of loosely woven or thick fabrics that fray easily.

The usual width of a finished bind is 0.6-1 cm but should be as narrow as possible on very fine fabrics.

General Method

1. Prepare crossway strips cut to four times finished width of bind (2.5-3.75 cm). Join one or more strips for the required length if necessary.

2. When used as an edging, trim off turning allowance on garment back to the fitting line as a bound edge neither adds nor takes away width from the cut edge. When used as a seam neatening trim turn to an even width (1.25-2 cm).
3. With R.S. facing, place edge of strip level with edge of fabric. Pin, tack and stitch 0.6 cm in from the edge. Remove tacks and press stitching.
4. Turn crossway up from stitching and press smoothly upwards. Turn to W.S. and without stretching the crossway, fold over the raw edge to almost meet edge of turning and press.
5. Bring folded edge over to the stitched line on W.S. and tack. Hem the fold on to the stitching. Remove tacks and press.

Binding Curved Edges

Follow the general method of binding with the following adjustments; the final result in each case should be flat, smooth and unplucked.

Inner Curves: armholes and neck lines

The edge of a concave or inward curve is shorter than the stitching line for the bind; therefore, slightly stretch the crossway when first pinning into position. Again stretch the folded edge before final hemming.

Outward Curves: shaped edges

The edge of a convex or outward curve is longer than the stitching line for the bind. Therefore, slightly ease the crossway when first pinning into position. Again slightly ease the folded edge before final hemming.

Binding Points

The binding of points and sharp angles is for decoration only, as a shaped facing would normally be used. Follow the general method of binding as given with the following adjustments.

Outward Points

Tack strip as far as the point and fasten thread

Ease the strip back around the point to form a triangular pleat. Tack across top of the pleat and continue.

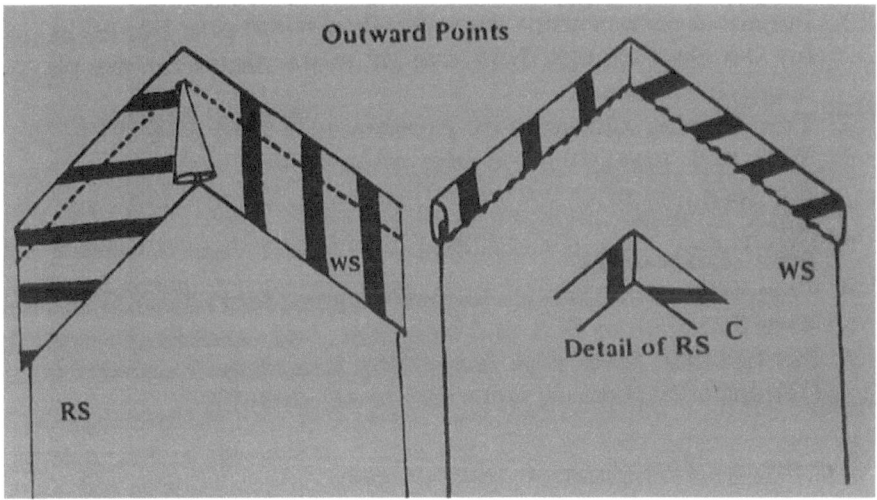

Stitch to the point and turn accurately, catching in top of pleat only.

On W.S. of point, ease in a similar small pleat to take up fullness making the point as sharp as possible and tack.

Hem into position on stitched line.

NOTE: S small fold therefore bisects the point on both R.S. and W.S.

Inward Points

Snip into the point across fabric of garment almost to stitching line. The inner angle can now open out into straight line.

Pin, tack and stitch binding into position as method given so that the stitching is close to the base of the snip which must be stitched securely.

With angle back in position, fold garment section in half for the point with R.S. facing. Backstitch across width of bind in line with the fold.

Open out flat and press this dart to one side and hem to the binding.

5.2.4 Crossway Strips Used for Facing

Unless required as a decoration, crossway facings are invisible from R.S. and are used to neaten armholes, necklines and hems and in the attaching of collars and cuffs. If a wide facing is required, it should be shaped facing.

General Method

Prepare crossway strips cut to finished width plus 1.25 cm to allow for 0.6 cm turnings. Join one or more pieces for the required length if necessary.

Trim turning allowance on garment to 0.6 cm from fitting line.

With R.S. together place edge of strip level with edge of fabric. Pin, tack and stitch 0.6 cm in from edge. Remove tacks and press stitching. Without stitching crossway fold up 0.6 cm turning and press.

Press seam turnings open to ensure a good fold line when finished. Turn facing on to W.S. and tack edge; fold carefully.

Pin and tack lower edge facing. Slip hem, then fold into position. Remove tacks and press carefully.

Crossway Facing as a Decoration

Use crossway strips of a contrasting colour applied to the R.S. of the garment. Follow general method as above, reversing position accordingly, thus at stage 3 place R.S. of facing to W.S. of garment. To finish, do not slip hem but work a line of top edge stitching as a decorative stitch.

NOTE: When bought 'bias binding' is used, the prepared folds are the stitching lines.

Facing Curved Edges
Follow the general method of facing with the following adjustments (it should be noted that these are the reverse to those for bound curves). The final result in each case should be flat and smooth.

Inward Curve: armholes and necklines
The edge stitching line of a concave or inward curve is shorter than the inner line of the facing. Therefore, slightly ease on the crossway when first pinning into position: slightly stretch the inner fold before hemming.

Outward Curve: shaped edges
The edge stitching line of a convex or outward curve is longer than inner line of the facing. Therefore, slightly stretch the crossway when first pinning into position. Slightly ease in the fold before slip hemming.

5.3 Interfacings
These are sections of firm, specifically prepared fabrics which are set into the garment to give added strength and support to areas of strain, e.g. full-length front openings with fastenings, waistbands and pockets. Interfacing also gives a firm and crisp tailored finish to collars, front openings, sleeve ends, cuffs, pocket vents and flaps.

Fabric available: from dress fabric and some haberdashery departments

Types of interfacings

Tailors canvas: various qualities and weights suitable for use with medium to heavyweight cloth for jackets, coats and heavier weight woolen dresses.

Muslin, scrim and tarlatan: medium and lightweight qualities suitable for dresses, blouses and lightweight jackets, some prepared as an iron-on fabric.

Bonded fabrics, i.e. non-woven: available in medium and light weights and also as 'iron-on' fabric. These non-woven fabrics re economical as they can be cut in any direction but they do not blend so smoothly with the main fabric as does woven interfacing. Superdrape is an interfacing which has special slits running in uniform lines allowing it to move with the fabric when stretched, i.e. for use in knitted fabrics such as jersey and woolens. A transparent interfacing is available for 'see through' fabrics, voiles, lawns, etc., which ensures that the interfacing remains transparent when ironed into place.

NOTE; 'Iron on' interfacings should be used with discretion as they reduce the pliable movement and handling of the main fabric although special interfacings are manufactured for knitted fabrics.

Three uses of interfacing are shown below. Interfacings must always be cut to the same grain as the section on which it is to be used and placed on the main fabric so that seam turnings with the added bulk fall towards the inside of the garment. For tailored collars, however, it is used on the undersection. The interfacings are basted in position on W.S. of the fabric before making up; it is then held in place by seams or construction lines. The basting is not removed until the garment is completed. Seam turnings of the interfacing must be trimmed to 0.3 cm to reduce bulk. No turning is to be taken into a seam.

1. Interfacing basted on to the under collar
2. Interfacing basted into position on the cuff edge of a sleeve
3. Interfacing basted into position on the bodice

5.4 Facing

A facing is a piece of fabric stitched that is shaped to match and fit the neckline, armhole or opening of a garment and stitched on the outside edge of a garment and folded over to conceal its raw edges. Facings are used to neaten the raw edges of and opening or to attach a collar. They can either be cut separately or extended from the garment.

Shaped facing

Shaped facings are cut similar to the area to be applied, they are used to finish armholes, necklines, openings and shaped edges. They are invisible except on coats and jackets and when they form a rever collar on dresses and blouses.

As a form of decoration they can be cut from contrasting fabric and applied to the R.S. of the garment.

These facings are cut as the pattern and grain of the garment but omit panel lines; the width and finished shape vary as is most suitable for the position and style. They are cut from the same fabric as the garment except where firm, thin fabric is more practical for use on thick or loosely woven fabric.

5.4.1 Facing on Armhole

All bodice seams must be completed.

With R.S. together join facing pieces on fitting line with a plain seam at shoulder and underarm. Press seams open and trim turnings to 1 cm.

Fold over 0.6 cm turning from the outer edge of the facing on to the W.S., and tack. Stitch 0.3 cm in from folded edge. Remove tacks and press.

With E.S. of facing to R.S. of the garment, match balance marks, fitting lines and seam lines at shoulder and under-arm. Pin, tack, and stitch into position. Trim seam turnings to 1 cm and snip into curved seam allowance, snipping more closely at the under-arm, where it is most curved.

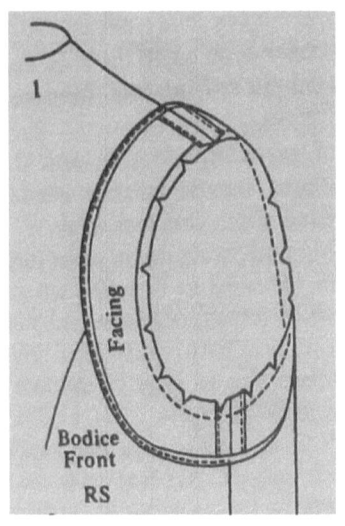 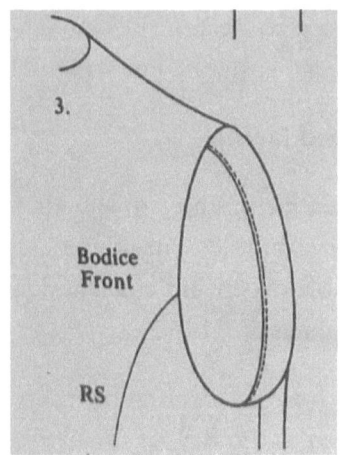

Turn facing through to W.S., bringing stitched line exactly on to the fold (or very slightly towards the inside) and tack this fold into position. Turn garment through to W.S. and press folded edge carefully.

Attach the facing in position by hemming on to the turning only of shoulder, under-arm and any panel line seams. Remove tacks. If the fabric is very resilient and the facings are likely to 'roll' at the folded edge, one of the following alternatives should be used.

NOTE: Before turning through to W.S., press facing up from the stitched line and against the turning. Stitch the facing on to the seam allowance only about 0.3 cm from stitched line.

Alternatively, turn facing through to W.S. and tack into position, and work a line of top stitching from the R.S. not more than 0.3 cm in from folded edge.

5.4.2 Facing a neck edge

The basic method is the same as that for facing an armhole. Instructions and diagrams are given for a neckline with centre back zip fastening; the same method would apply for a front neck opening. Before facing the neck, shoulder and panel lines, seams should be neatened and the opening completed.

1. With R.S. together and matching fitting lines, join front and back neck facing with a plain seam. Press seams open and trim turnings to 1 cm.
2. Fold over 0.6 cm turning from outer edge on to the W.S. and tack (not C.B. edge). Machine stitch 0.3 cm in from fold. Remove tacks and press.
3. With R.S. of facing to R.S. of garment, pin C.F. and C.B. in position. Then, matching shoulder seams, balance marks and fitting lines, pin, tack and machine stitch the neck edge.
4. Remove tacks and press stitching and using a pad press turnings open to ensure a good fold. Trim turnings to 1 cm (0.6 cm on fine fabrics) and snip seam allowance of the curved edge and trim off corners at C.B.
5. Turn facing to W.S., bring stitched line to edge of the fold and tack into position.
6. Turn garment through to the W.S. and press edge fold carefully. Pin shoulder seams in place. Turn under C.B. edges on to the tape of the zip fastener and hem. Hem facing to turnings of shoulder and panel line seams. Remove tacks.

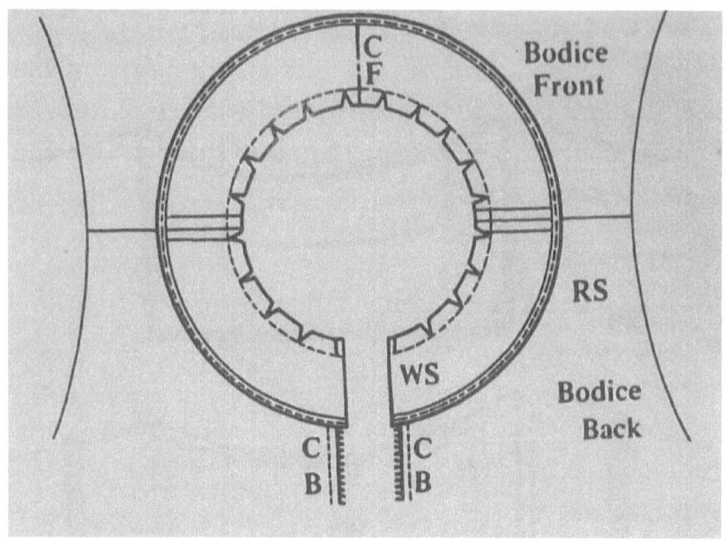

5.4.3 Facing a square neck

Follow the instructions given for facing a neck and armhole as appropriate.

To attain a square finish tack the neckline carefully and stitch the corners accurately. After trimming the seam allowance snip diagonally into the corners and around the curved edge as shown in the diagram below.

The square corners are weak and it is advisable to work a line of edge stitching from R.S. after the fold has been tacked in place, as in the diagrams below.

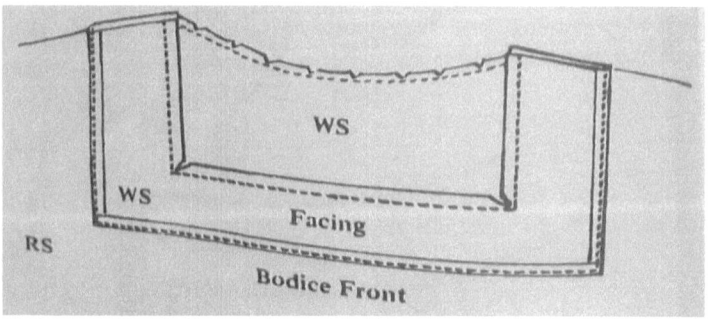

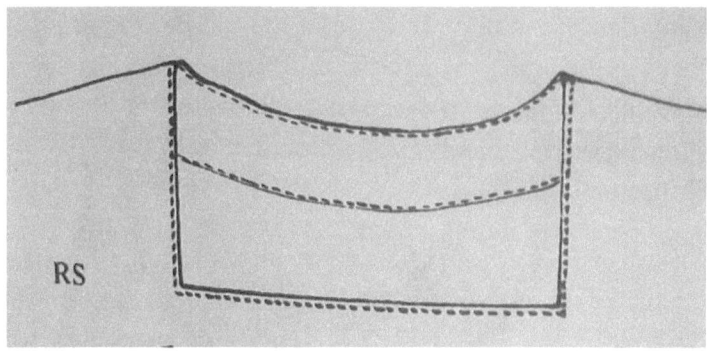

5.4.4 Decorative neck facing

Follow the basic instructions with these adjustments at:

Stage 2: Tack but do not stitch edge folded on to W.S. of facing.

Stage 3: Place the R.S. of the facing to the W.S. of the garment. Continue as directed.

Stage 5: Turn the facing on R.S. of garment and tack fold in place. Pin seams and points correctly, then carefully pin and tack edge of facing into position.

Finish both neck edge and facing edge with a line of top stitching 0.15 cm in from the fold. Remove tacks and press. A decorative stitch can be worked instead of final top stitching

5.5 Linings

Linings are stitched into the wrong side a garment primarily to prevent opaqueness and make the garment easy to wear. It will also prevent the garment from sticking to the body. Lining will also increase the durability of a garment.

Lining a skirt

Cut the lining patterns from the skirt patterns, sew the lining patterns, leaving a gap for the zip. Darts will not be stitched.

1. Pin the lining to the skirt at the waist. Match all the seams.
2. Make pleats in the lining to match to the darts on the skirt.
3. Machine stitch the lining to the skirt at the waist line together with the facing or waistband.

Hemming a lining

The lining on a skirt or dress should be slightly shorter about 4 cm (1 ½") than the finished garment, this will prevent the lining from showing while walking.

1. Machine the hem on the lining, making a double turn hem of 4 cm (1 ½").
2. Hem the garment in place. Make sure the lining hem should is about 4 cm from the hem fold.

5.6 Revision Questions
1. Define crossway strips
2. List areas where crossway strips are applied
3. Define interfacing
4. List the types of facings
5. List the types of interfacings
6. List areas where interfacings are applies

Chapter 6: Openings

6.1 Introduction

This chapter explains the major openings and the types of fastenings used on them. Small openings with simple fasteners are very essential to Blouses, lightweight dresses, lingerie and children's wear. The opening must be long enough for ease in dressing with the end accurately stitched strongly to withstand strain.

Openings can be planned to be both practical and decorative. The choice of opening is determined by;

    a. Position of the opening
    b. Purpose of the opening
    c. The type of fabric used

The three types of opening are;

Faced opening: simple, unobtrusive or decorative

Bound opening: visible, contrasting fabric may be used

Continuous strip opening: string with underlap

6.2 Faced opening

Faced openings are desirable for front and back neck openings and wrist openings on sleeves.

1. Mark the position and length of opening on the garment with a line of tacking
2. Cut out the facing: length of the opening plus 5 to 6.25 cm.
3. Mark out the length of the opening with tacking down the centre of the facing.
4. Neaten three raw edges with edge stitching
5. With R.S. together, place the two tacked lines directly on top of each other and tack together.

6. Starting 0.6 cm out from the tacked line, machine stitch down to the base of the line end up again 0.6 cm out at the top

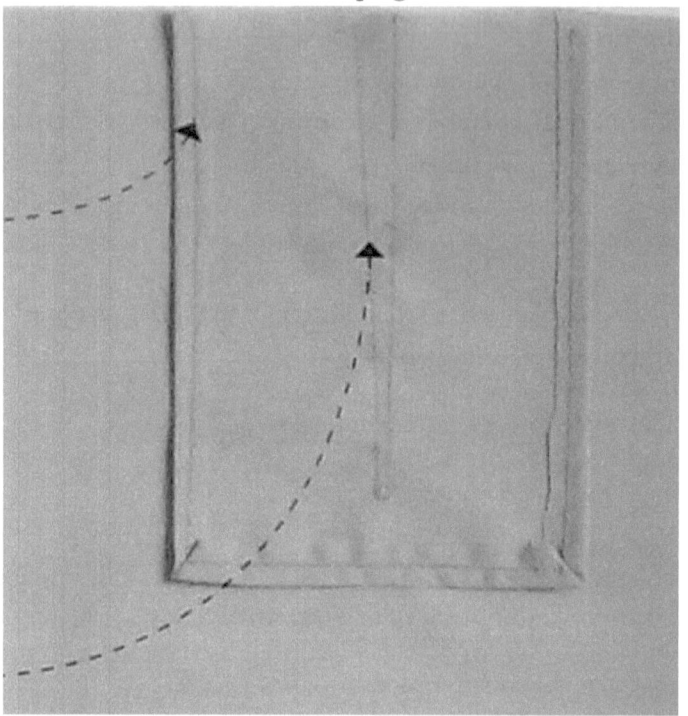

7. Cut down the tacked line as far into the point as possible.

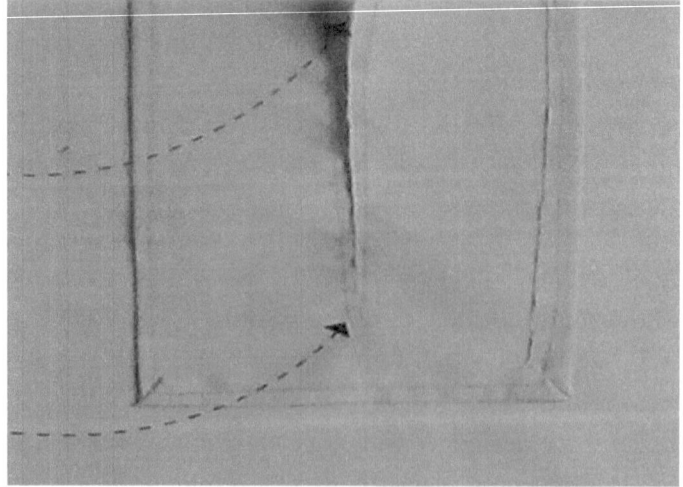

8. Turn the facing through to W.S. of the garment.
9. Tack the folded edge into position and press

10. Strengthen the opening by edge-stitching from R.S.

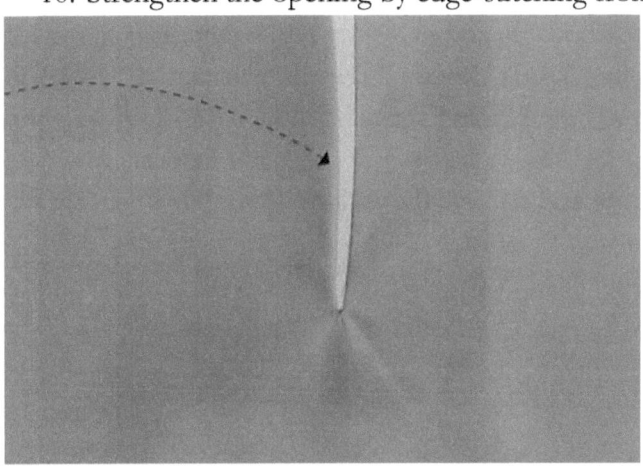

6.3 Bound opening

This opening is used for front and back neck openings. A crossway strip that contrasts with the garment can be used to make the opening decorative

1. Cut the required length of the opening

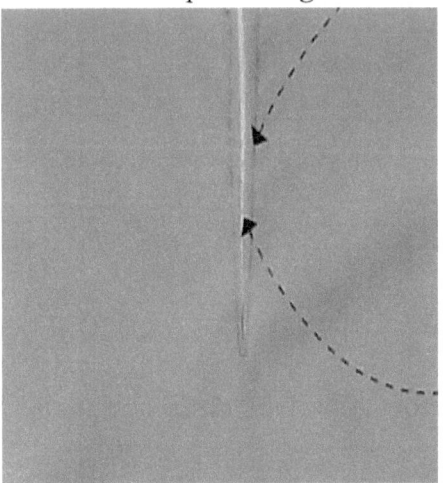

2. Cut a crossway strip twice the length of the opening and 2.5 cm wide.

3. Open out the opening and place the R.S. of the strip to the R.S. of the opening. Tack to within 2 cm of the base.
4. Slightly stretch the binding for the following 3.75 cm and then complete the tacking as before
5. Machine stitch and remove the tacking

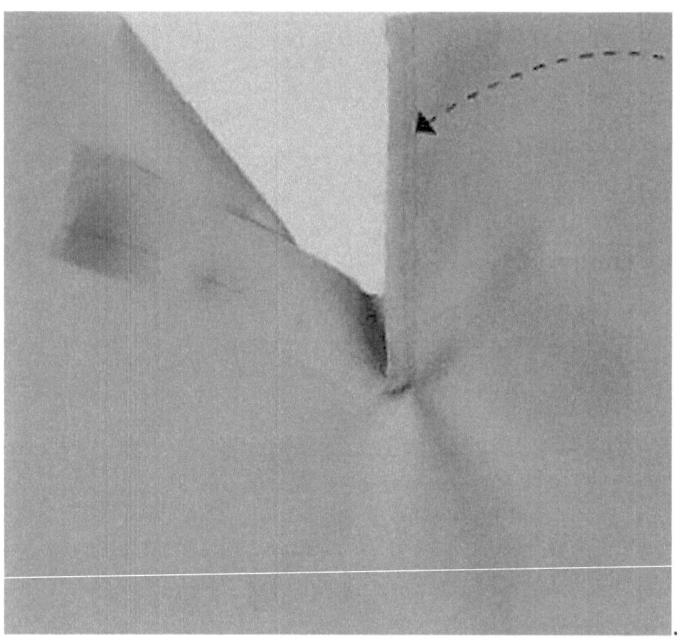

6. Make 0.6 cm turning on to the W.S. of the strip.
7. Fold over the strip so that the fold touches the machine stitching.
8. Hem through the machine stitching so that the stitches do not show on the R.S.
9. Stretch the binding as before at the base of the opening.

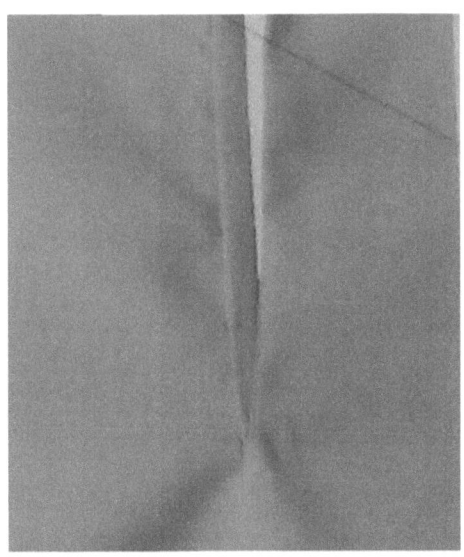

6.4 Continuous strip opening

Continuous strip opening is suitable for use in lingerie

Cut a straight slit the length of the opening required

Cut a straight of grain strip twice the length of the opening and 5-6.25 cm wide.

Open out the slit with the W.S. facing the worker.

Place the R.S. of the strip to the R.S. of the opening.

Pin the strip into position with the garment turning tapering to almost nothing at the base of the opening.

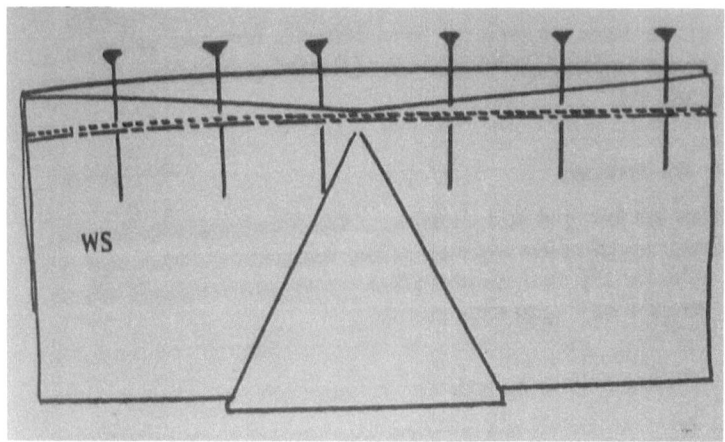

Tack and machine stitch, with 0.6 cm turning parallel with the edge of the strip. The stitching needs care where the garment turning is narrow at the base. It must be smooth and unpuckered.

Remove the tacking and press the strip and turnings away from the garment. Crease 0.6 cm turning along the free edge of the strip.

Fold over the strip, pin and tack into place so that the fold touches the machine stitching.

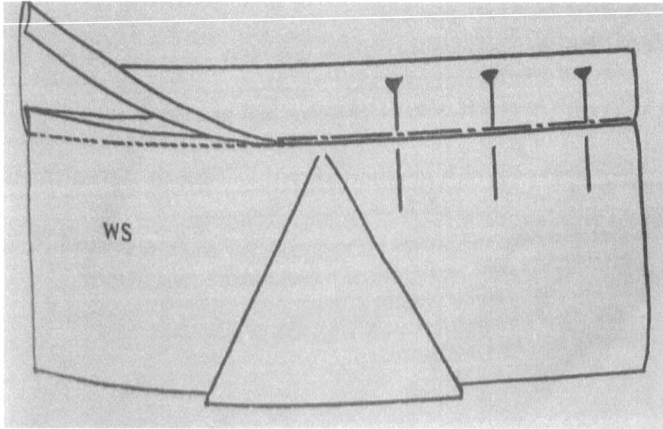

Hemstitch the edge of the strip into the machine stitching. Remove the tackings and press.

Fold the strip into place to form under and overlap and backstitch 0.3 cm from the base of the opening to strengthen it.

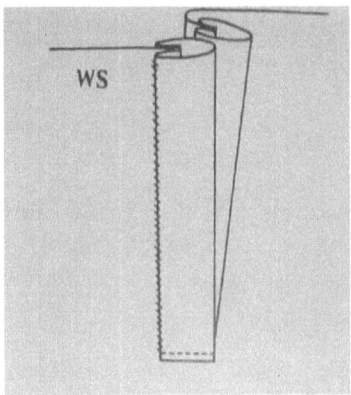

Placket opening

Plackets are neatened vents or faced openings designed particular areas of garments. Examples of garments where placket are applied are bodice, skirt, sleeve, jacket, dress, and pant. Plackets can be finished in any style rounded, pointed, stylized or blunt ends length and width may also vary. Plackets may also be fastened with buttons and buttonholes or not.

6.5 Revision Questions
1. List the three main types of openings
2. State how you would make a bound opening decorative
3. Outline areas where faced openings are best suitable in a garment
4. List the five types of garments where placket openings are applied.

Chapter 7: Fastenings

7.1 Introduction

Fastenings are objects sewn or attached into a garment that allow to be fastened. There are different methods of fastening an opening.

Zip Fasteners

Zips are firm and neat fasteners that form a continuous edge-to-edge fastening. They are quick to fasten and can be sewn to be visual or inconspicuous. Zips are used for top clothes only; they are available in various weights of according to requirements.

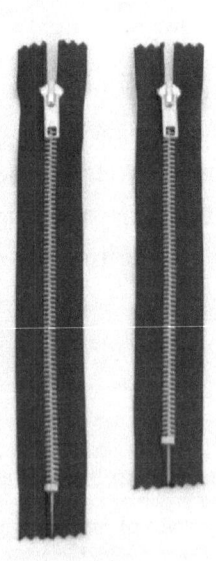

Button, Buttonholes

These are visual methods of fastening unless when used in a fly opening. Buttons are not applicable on areas of strain and can be considered as part of the style of the garment and its decoration.

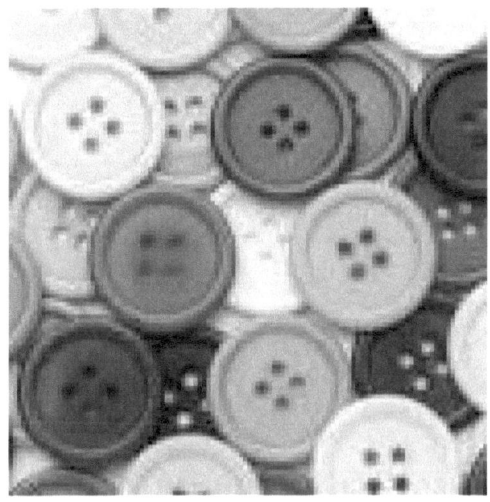

Button and button holes are used on all garments.

Press Studs:

Press studs are firm and secure method of fastening openings with little strain and are inconspicuous. They are used for shoulder and neck openings on young children's cloths. They can also be used on wrist and similar openings and as a finish for a main buttoned opening.

Hooks and Eyes or Bars: these are firm and secure under strain. These types of openings necessitate strain in order that the hook and eye remains fastened. They are inconspicuous and are available in varied weights according to requirement. Use for waistbands and as a finishing detail to a zip fastener or a buttoned opening.

General Rules for Fastenings

- Fastening must always be sewn on double fabric
- Fastenings must be inconspicuous unless the fastening is to be decorative.
- Particular care and attention must be given to detail, accurate stitching and secure fastening on and off of the sewing thread.

7.2 Zip fasteners

The type and weight for the opening is chosen by considering the weight of the fabric. The colour of the tape should match that of the garment unless a contrast is required for decoration. Types available are

Featherweight: for neck and wrist opening on light fabrics

Lightweight: for dresses, cotton skirts and shorts of medium weight fabrics

Skirt weight: for skirts, shorts or trousers on suiting materials

Open ended: for jackets, cardigans and coats and heavy fabrics

Curved: for front openings on trousers

Invisible: for skirts and dresses

Decorative: for dresses and jackets

There are four methods of inserting a zip fastener:

Visual: for decoration only zip will be visible on the right side.

Semi-concealed: for neck, wrist, side and C.B. openings in lightweight fabrics for open-ended zip.

Concealed: for side C.F. and C.B. openings in dresses, skirts, shorts and trousers.

Invisible: for invisible fasteners only.

7.2.1 Visual Method: for inserting a zip into a panel without a seam

For a position where the zip forms part of the decoration, e.g. C.F. or pocket fastening.

1. Complete a faced opening following the method but instead of making a point, finish with a square end. The width of opening should be the length of zip teeth from fitting line.

2. Working from the R.S. of garment with zip closed and the top of the slider level with the fitting line, tack the zip into position, keeping the edge of the opening parallel with the teeth.
3. Attach the piping foot to the sewing machine and, working from R.S., stitch the faced opening on to the zip tape, making sure that the end corners are square
4. Turn to W.S. and oversew together the inner edges of the tape at the base of the teeth. Hem the side of the tape to the facing and loop stitch across the raw edge at base.
5. Hold the lower edges of the facing in position by slip hemming on to the garment.

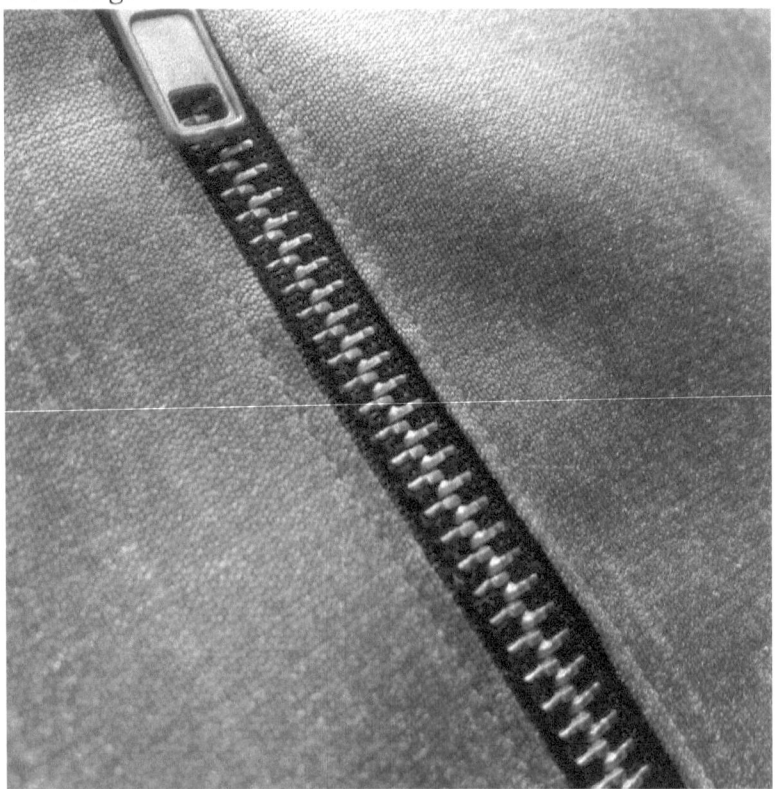

7.2.3 Semi-concealed Method: for inserting a zip into a plain seam

For positions with C.F., C.B. and side openings of dresses and over-blouses; C.B. openings on skirts and open-ended zips on jackets and coats.

1. Prepare and stitch the plain seam as far as the opening which should be the length of the zip teeth from the fitting line. Press the seam allowance flat so that the fitting line remains on the fold of open section. Neaten the raw edges of both seam and opening.
2. With a marker, tack a straight line along each side of opening 0.6 cm in from the folded edge and remove tacks.

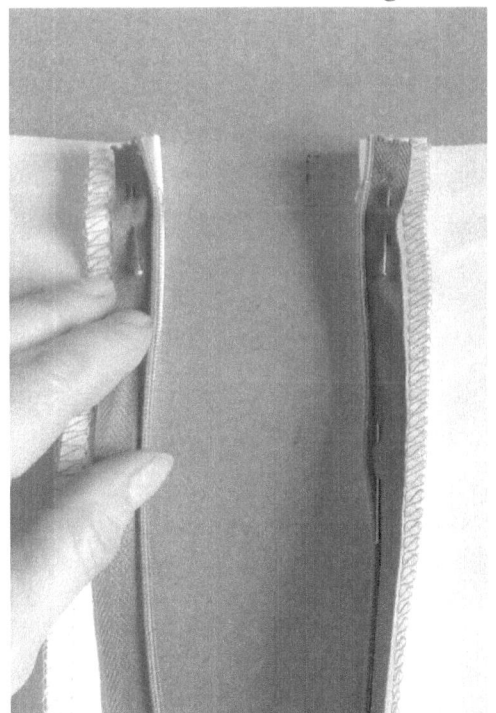

3. Working from R.S. with the zip closed and the top edge of the slider level with the fitting line, place the right-hand side of the opening over the zip so that the folded edge is level with the centre of the teeth and tack into position. On a C.B. opening it

is advisable to have a small hook and eye at neck edges; therefore, set zip 0.3 cm down from fitting line.

4. Hand stitch zip in position by working a back stitch over every third stitch of the machine stitched line. Hand stitch as before long the machine line and backstitch across the base to give a square finish.

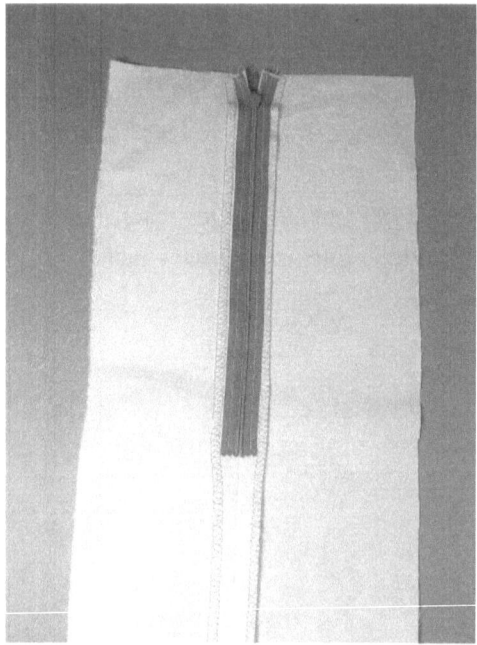

5. Tack the left-hand side in place so that the two edges meet and conceal the zip. Hand stitch as before along the machine line and backstitch across the base to give a square finish.
6. Neaten the tape edges on the W.S. as for the visual method. Attach hook and eye at neck edge on C.B. opening.

7.2.3 Concealed Method: for inserting a zip into a plain seam

For positions with C.F., C.B. and side openings of dresses, skirts, shorts and trousers, and for open-ended zips on jackets and coats.

Prepare and stitch the plain seam as far as the opening which should be the length of zip teeth from the fitting line. Press the stitched seam allowance flat and also the seam allowance of the front section only of the opening so that the fitting line is on the fold.

On the back section of the opening, fold back the seam allowance 0.3 cm beyond the fitting line. To allow the seam to lie flat, snip across seam turning just below the base of opening. Machine stitch the edge of the fold; neaten the raw edge of the seam opening and snipped turning.

Working from R.S. of the garment with the zip closed and the top edge of the slider level with the fitting line, place the back section of the opening over the zip tape so that the folded edge is against the edge of the teeth: tack into position. Hand stitch over the machine line.

To enclose the zip, tack the front overlap into place so that the edge fold is level with the fitting line on the back section. Hand stitch over the machined line and backstitch across the base to give a square finish.

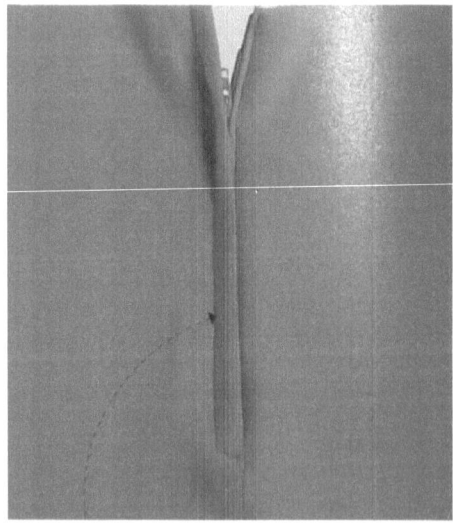

7.2.4 Invisible Method

A method used for inserting a zip into a plain seam where neither the zip nor stitching is visible on the right side. These zips are of a different construction from the ordinary zip fastener: they are applied on centre front, centre back and side openings on dresses and over-blouses and side and centre back openings on skirts, shorts and trousers.

1. Pin and tack the plain seam including the opening. Stitch the seam to 1.25 cm above the base position of the zip: therefore, the opening will be the length of zip teeth less 1.25 cm from fitting line. Leave the opening section tacked but remove tacks from the stitched seam. Press the turnings open and neaten the raw edges.
2. Attach a piping or zipper foot to the sewing machine. Adjust the machine to approximately ten stitches to 2.5 cm.
3. Working from W.S. of the garment with the zip closed and the top of the slider level with the fitting line, place the zip centrally over the tacked seam line. Tack the tape
4. Firmly to the seam allowance only. On C.B. neck openings, position the slider 1.25 cm below the fitting line.

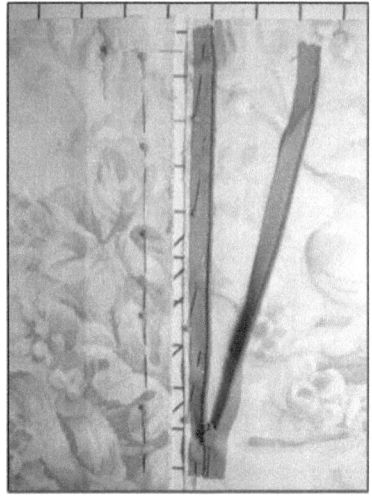

5. Remove the seam tackings and open the zip fully.
6. Place the whole of the garment to the left-hand side so that the right-hand side of seam allowance can be flat. With the teeth opened back, machine stitch as close to the teeth as possible

through the tape and seam allowance only, working from the top of the opening to the top of the zip slider. Fasten off.
7. Reverse the garment so that the left-hand allowance can lie flat and, with the teeth opened back, machine stitch the second side from the top of zip slider to the top end of the opening.
8. Gently and carefully close the zip fastener.
9. Position the garment as at stage 5 and, stitching through the tape and seam allowance only, stitch from 1.25 cm above the base of the zip to the end of the tape. Reverse position of garment and stitch the second side.
10. Neaten the tapes as in the method given. Attach hook and eye at neck edge of C.B. opening.

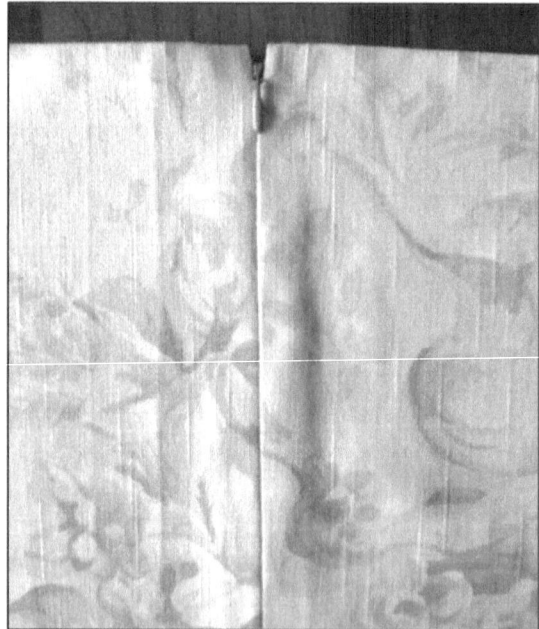

7.3 Fastening with Buttons

There are four kinds of buttonholes:

Worked buttonholes: generally suitable for most fabrics and openings.

Bound buttonholes: for use on top garments of heavier fabrics only.

Buttonhole loops: these are usually for small openings on children's wear, wrist or neck openings, all fastened with a single button. They can also be used with a single button that lies under the collar of a main opening on blouses and coats.

Rouleau loops: a decorative method used with several small covered buttons and occasionally on top coats.

Position of hand or worked buttonholes

These are made on the extension over section of the garment opposite the buttons on the underlap. They are placed so as when fastened, button lie on centre front or centre back line and at the centre of all openings.

Direction: cut along the straight thread in the direction of the strain; therefore button holes are horizontal on waistbands, cuffs and main openings. On the shoulder straps, shoulder and crutch openings on children's wear the button holes ate straight in appearance but are cut across the opening to take the strain and are regarded as horizontal buttonholes and are worked accordingly. On front openings of skirts, shirt blouse and light dresses where there is very little strain, vertical buttonholes are worked

Length of slit: equals the diameter of the button plus 0.3cm

Overlap: this must extend beyond the C.F or C.B line at least one half the diameter of the button plus 0.3cm. Overlap extension and method of marking is the same foe all other buttoned openings.

Marking the position for horizontal buttonholes:

1. Tack in C.F. or C.B. line. Tack a line parallel to this the diameter of the button inside from the centerline
2. Mark the position of the buttonhole slit to correspond with the center of button placing, tack on the straight thread between

parallel lines and extend 0.3cm beyond the centerline. This ensures that the button will rest on the centerline.

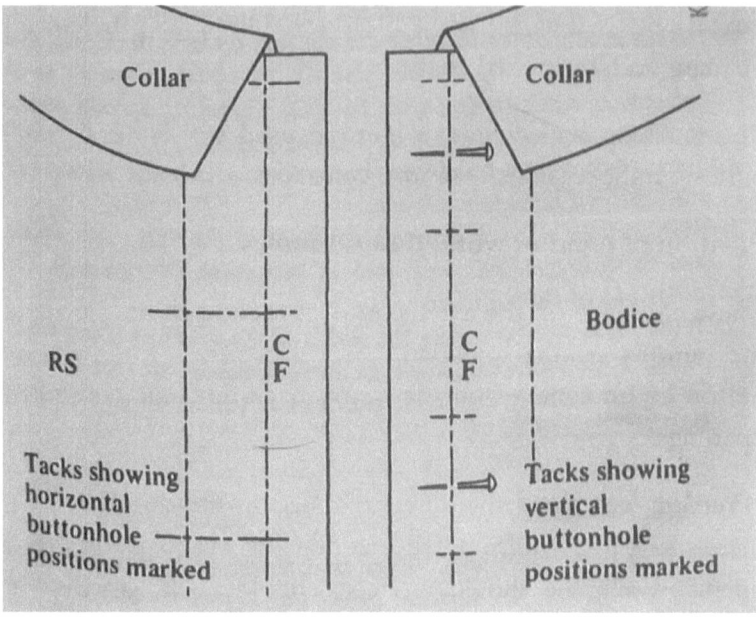

Marking the position for vertical buttonholes

1. Tack in C.F. or C.B. line and mark with a pin the position to correspond with the centre of the button.
2. Tack mark half the length of the buttonhole slit on either side of the pin.

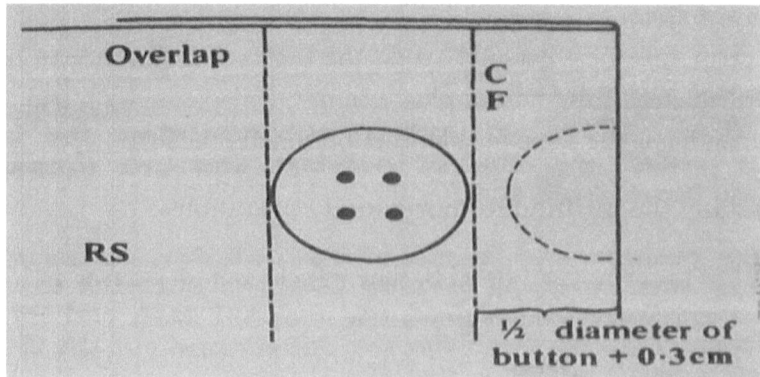

7.3.1 Worked buttonholes

Worked button holes are used on all weights of fabric. Worked buttonholes are made on two layers of fabric, button holes are made after facings, collars and cuffs have been completed.

Horizontal buttonholes

These have one round end to accommodate the button shank, worked nearest the edge of overlap, and one square end to give strength. Work from R.S.

1. Mark position and length of buttonholes; work a line of small running stitches (stabbed through thick fabric) 0.15cm away from tacked line and around each end
2. Snip carefully through the two layers and cut the exact length of the cut mark, along a straight thread.

On loose fabric oversew the raw edges with a fine thread to prevent fraying.

3. Work buttonhole stitch from left to right starting at lower inside corner of slit and continue to end of slit.
4. Taking needle down through the slit and up into the fabric, oversew the round end, stitch should be equal in length to the buttonhole stitch in line with the slit. Therefore an odd number of stitches is used, either 5 or 7.
5. Turn work in the hand and buttonhole stitch the second side, still work from left to right.
6. Pass the needle down through the knot of the first stitch worked and bring it out level with end of last stitch. This holds end of slit together.
7. Work three stitches across the width of the stitches. Work buttonholes stitch across these threads, so that first and last stitches are in line with previous stitching, to give a square finish.

Vertical buttonholes

These can be worked with either two round ends in which case follow the method given but complete with a second round end of oversewing or with two square ends, in this case work both sides first then work square bar at each end.

7.3.2 Bound buttonholes

Bound buttons are used on medium and heavyweight top garments and can be made on cross or on bias where necessary in which case a piece of tape must be tacked to the wrong side during preparation.

This is not a suitable method for buttonholes that will take great strain, such as waistbands or straps on children's wear.

Bound buttonholes are made first through single fabric ONLY (or fabric and interfacing when used), they are not completed until the facing has been attached and is in its final position.

Working from Right Side

1. Mark position and length of buttonhole through single layer of fabric. Cut crossway binding 5cm wide by length of buttonhole plus 2.5cm. Fold the strip in half lengthwise and press. Lay the fold against the position and tack with R.S. facing.
2. Open up strip and tack through both thickness along the creased line. Work a rectangle of machine stitching 0.3 – 0.5 cm from tacked line starting one-third of the way along the long side, and marking the same number of stitches across each end; finishing by work over 3 – 4 of the first stitches.

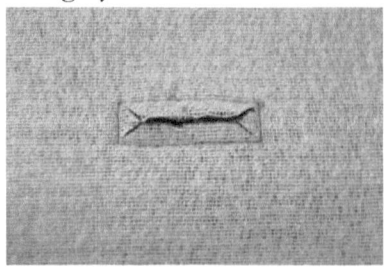

3. Remove tack marks and press. Fold buttonhole back in half and snip through each layer, open out and cut to 0.3cm from each end and snip into corner.

4. Pass the binding strip through to W.S.
5. Pull the strip out flat and press. This ensures that the corners are snipped sufficiently so that strip can lie flat without puckers.

6. Place the turnings back to their original position, i.e. flat, and fold the binding strip back over them on each long side, to form an inverted pleat at each end of slit. Over-sew the folded edges outside the buttonhole slit. Over-sew the folded edges outside the buttonhole slit. Tack the binding into position along the machine stitched line. If fabric is difficult to handle, stab stitch through to R.S.

7. When facing has been attached and folded into position to cover the back of the buttonholes, baste to hold in place. Cut a

slit in the facing to correspond with length of the buttonholes. Turn the raw edges under with a needle and hem the facing around the bound buttonhole. Remove tacks and press.

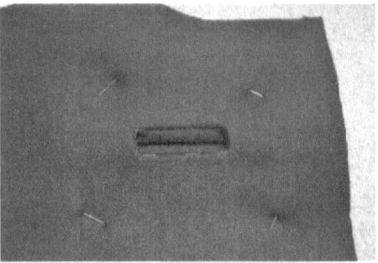

7.3.3 Worked loops

For use on children's wear and small openings that meet edge to edge, but do not overlap. Also for use with the small under collar button on garments with a main front opening that button up to the neck.

Use buttonhole twist. Complete neatening and garment first. Work from the R.S.

1. Place a piece of thin card behind the position of the button loop and hold in place with two pins, diameter of button apart.
2. Place a third pin so that it enters the card half diameter of button away from edge of opening and bring pin up towards the edge.
3. Securely fasten a thread and passing behind the third pin, so that thread rests, make stitch across fold. Return in the same way, and back again, making three loops in all, work a further small stitch to hold thread.

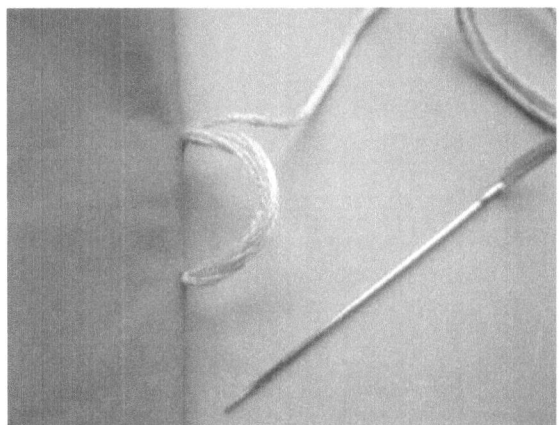

4. Remove pins and card. Using the thread work loop stitch over the strands to finish. Stitches should be close and firm with the looped edge outwards.

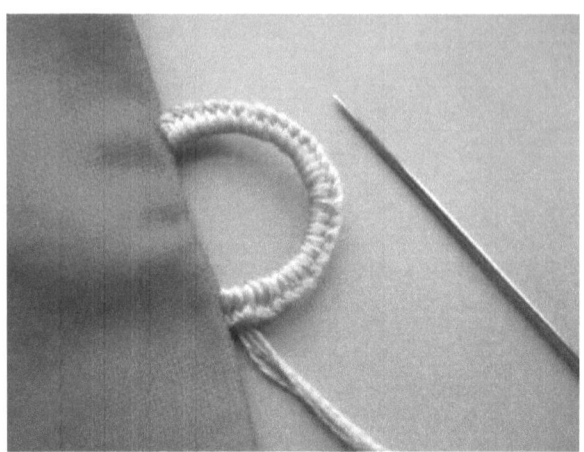

7.3.4 Rouleau loops

Decorative method of fastening on edge to edge opening at C.B., C.F. or at wrists. Buttons are small, self-covered in fabric and are arranged almost touching in either groups or in a continuous line. (If required, spaced buttons mat be sewn on an underlap to prevent gaping.)

Rouleau button loops are made before opening is faced. Work from R.S.

1. Make sufficient length of Rouleau.

2. Tack the fitting line and tack a second line parallel, the diameter of button apart. Mark the position of button
3. Curve the Rouleau to form loops with the seam inside the curve and tack into place, so that the outer edge of loop touches the width tack and allowing 0.6 cm turnings beyond fitting line and machine stitch the loops along fitting line.

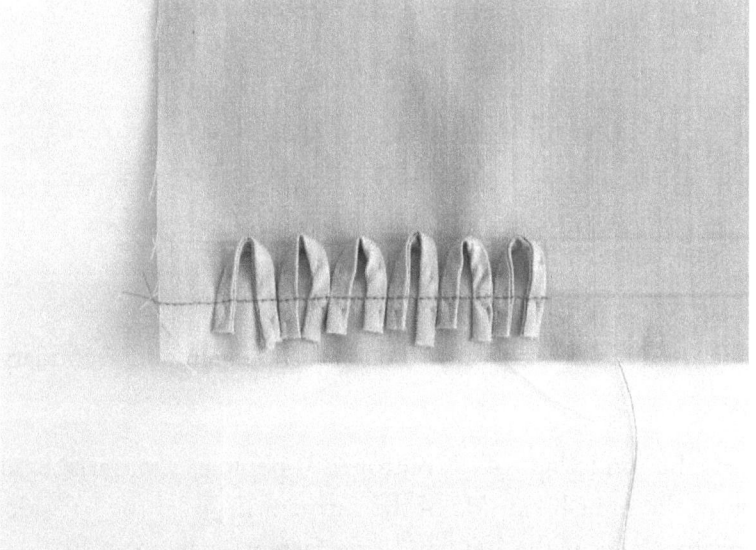

4. Place the facing over the loops along fitting line with R.S. together, pin tack and machine on fitting line. Remove tack and press stitching. Trim turnings to 0.6 cm.
5. Turn facing back on to W.S. so that loops protrude from the folded edge.

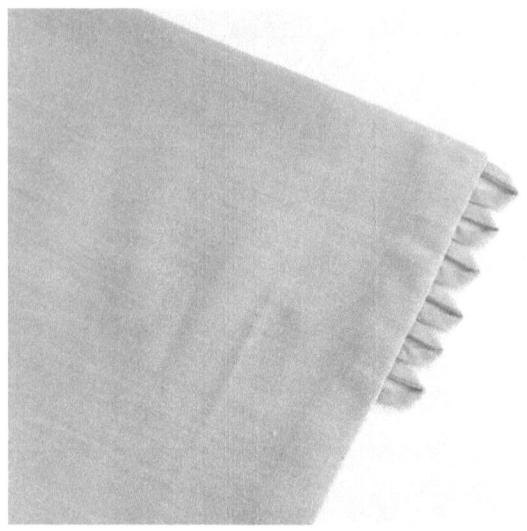

7.3.5 Buttons

These can be used for decoration and all types of main and secondary fastenings.

Position: on the underlap of a buttoned opening. On girls' and women's wear, the right-hand side of the garment fits over the left side: on boys' and men's wear the left-hand side fastens over the right side. Shoulder and crutch openings fasten front over back.

Buttons are placed centrally on the C.F. or C.B. lines and on all other openings. Enough buttons should be evenly equally to make the opening look neat when closed, with no gaping between buttons. The distance of spacing depends on the position of opening, weight of the fabric, and size of button.

For decoration, buttons can be (a) paired ; (b) sewn on both sides to give a double-breasted impression; (c) on either side of an edge-to-edge front opening fastened with either a conspicuous worked loop or a decorative braiding; and (d) as being purely decorative and not as a fastening.

Sewing on buttons

Buttons must always be sewn on to double fabric; where the fabric is single reinforce with tape or interfacing hemmed to W.S.

A shank between button and fabric must be in line with the buttonhole.

Four-hole buttons may be stitched.

Attach buttons after buttonholes have been worked.

1. Mark the position of each button to correspond with the buttonholes. (Button position usually marked on trade patterns.)
2. Using a double thread of matching colour fasten on securely. Work five or six upright stitches through buttonholes leaving 0.3 cm thread each time between button and fabric to form a shank. On thick fabric it will be necessary to allow thread for a longer shank to accommodate thickness of buttonhole. When a button has a shank already on the underside allow for a further thread shank as necessary.

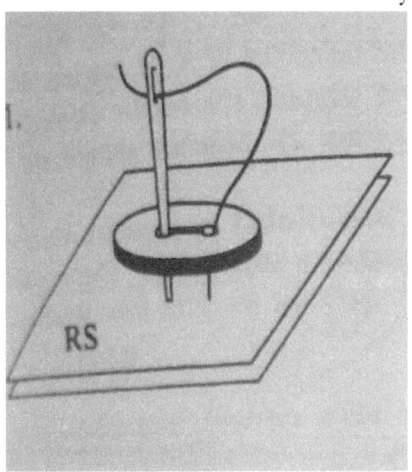
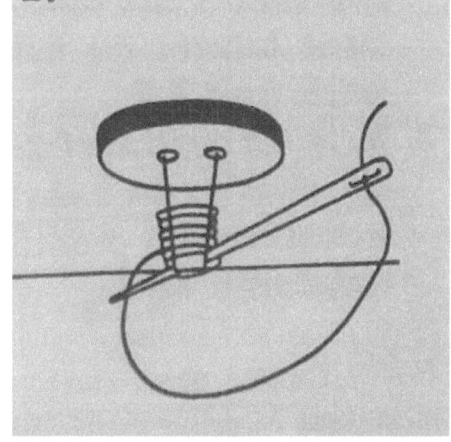

7.4 Press Studs

Used for openings with little strain and as a supplement to a main buttoned opening to control corner of overlap.

The section of press stud with a knob is sewn on to the overwrap so that it may be fastened by pressing down into the section with a central hole, that is sewn on the underlap. Use thread that matches the colour of garments.

1. Sew the top part of the press stud on to the W.S. of overlap. Work four stitches into each hole and slipping the needle through the fabric between the holes. Stitches should not be visible from R.S.
2. Mark the position of the underpart of stud by pressing the knob into place on the underlap to make an impression. Place undersection centrally over this mark and sew on with four stitches in each hole as before.

NOTE: Larger press studs may be sewn on with buttonhole stitch using buttonhole twist.

7.5 Hooks and eyes

For use in positions of strain as a degree of strain is necessary for hook to remain fastened.

Particularly useful for waistbands and as a secure finish at neck edge of a zip fastener.

Hooks: Available in large to very small size in black and white metal, supplied with either curved or straight metal bars.

1. Fasten thread with a double stitch, where it will be hidden by the hook.
2. Place bend of hook 0.3 cm in from edge of opening and sew in position with 3 straight stitches under the actual hook.
3. Slip needle through fabric to the ring and work buttonhole around each ring.

4. Finish with 3 straight stitches across centre of hook between rings and 'hump' under the hook. Fasten off.

NOTE: Special skirt hooks, flat and strong, are available for waistbands of skirts, shorts and trousers.

Curved metal eyes: for use on edge-to-edge fastening. The loop must extend 0.3 cm beyond edge of opening, placed to correspond with hook. Work buttonhole around each ring and three straight stitches across centre.

Straight metal eyes: for use on overlapped fastening. Bar is placed 0.3 cm beyond fitting line to correspond with hook. Work is buttonhole stitch around each ring.

Worked bar: neat and less conspicuous, used in place of metal eyes with small or very small hooks. Bar is made either on folded edge or on fitting line.

Work a loop stitched bar over three straight stitches, with required to take hook.

7.7 Revision Questions

1. List groups of fasteners
2. Explain the rules for fastenings
3. List the methods of inserting a zip
4. List the types of buttonholes
5. Outline where open ended zips are applied
6. List the factors that determine the width of spacing of buttons on a garment.

Chapter 8: Pockets and Collars

8.1 Pockets

It's recommended that a pocket is strongly made and large enough for its purpose. It must be strongly attached to avoid strain on the garment. When visible on the right side pockets are both decorative and functional. But become functional only when concealed in a seam.

Two general methods are shown.

a. Patch pockets – stitched onto the right side of the garment thus entirely visible
b. Welt pockets – inserted into the garment thus only the opening is visible

8.1.1 Patch pockets

The pocket is made by top stitching a piece 'patch' of fabric on to the right side of the garment. The pocket is used to carry small articles and can also be used for decoration.

Methods of neatening the top edge of the patch pocket:

(a) Turning a plain or shaped hem to the R.S. and top stitch it in place
(b) Attaching a straight or shaped facing of self-fabric on or of a contrasting decorative fabric

To attach pocket:
1. Complete the top edge as required.
2. Fold turnings of pocket to W.S. on fitting line and tack. Trim turnings to 0.6cm. on square corners and points pleat and on curved edges snip turnings to reduce bulk

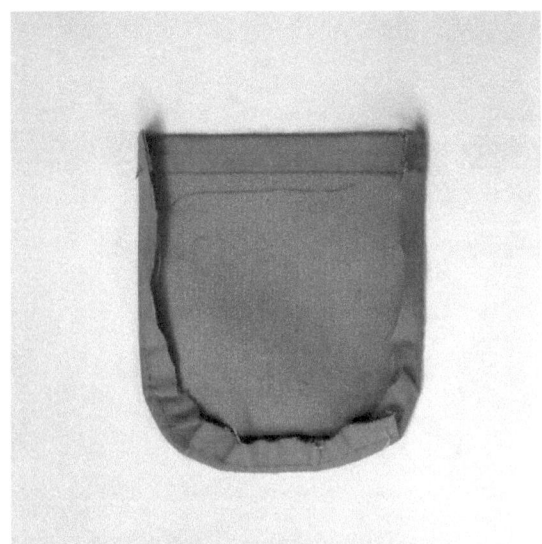

3. Match balance marks, place the pocket in position on the garment and tack.

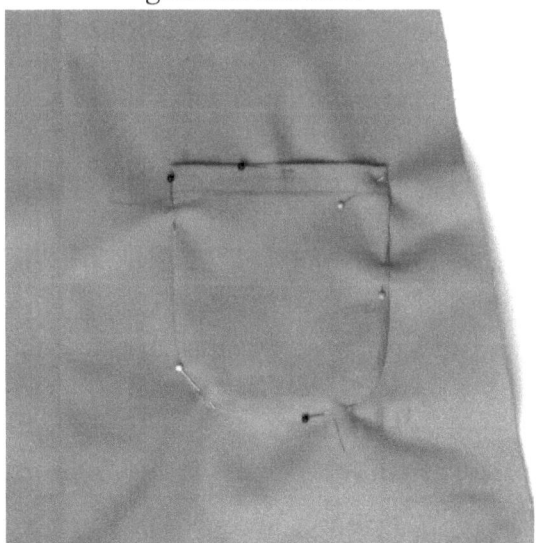

4. Machine stitch by working support of corner first, proceed around edge of pocket to finish with the support of second corner.

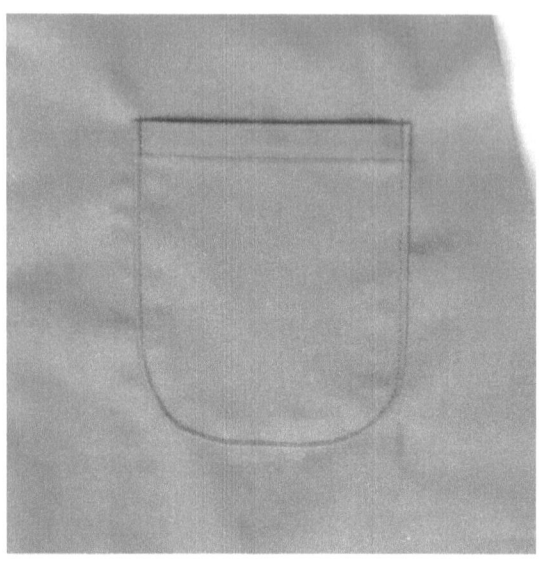

8.1.2 Welt Pockets

This is a pocket that is inserted into the garment with the opening strengthened by a welt. The welt is the only visible section of a welt pocket when completed. The pocket can be inserted on a straight grain or at an angle to it. The pocket bag is made from the same fabric as the garment to form the back and the front cut from a lining fabric.

1. Mark the position of pocket mouth, if not on the straight grain, baste a piece of tape on interfacing to the W.S. pocket mouth should be at least 9cm wide.
2. Fold the welt in half lengthwise R.S. together, tack and machine stitch the ends on the fitting line. Press, trim and turn R.S. out press again.
3. Tack the prepared welt into position on the R.S. of the garment so that welt faces downwards and the fitting line is 0.5cm below the position mark for the pocket mouth.
4. Place the front lining section of the pocket over the welt, matching fitting lines. R.S. of lining to R.S. of garment, tack through all layers. Place the back section of the pocket above the pocket mouth with R.S.

together and tack into position with fitting line 0.5cm above the position mark for the pocket mouth.
5. Machine stitch on the fitting line each side to exact width of pocket mouth. Ensure that lines are parallel and equal in length. Fasten off all ends very securely. Remove tacks and press.
6. Fold the pocket mouth in half and snip on the fold, open out and cut to within 0.6 cm of the ends and then diagonally into the corners. Turn both pieces of the pocket through to W.S. and press seam so that each piece hangs downward. Press the welt into an upright position on the R.S. to hide the pocket mouth.
7. Tack the front and back of the pocket together making them the same width as the welt. Machine stitch along the tack line. Press, trim the turnings to 0.6 cm and neaten by oversewing.
8. Tack welt into position of the R.S. of the garment and blind hem or over-sew both ends from R.S. making both top points very strong as they take the most strain.

8.1.3 Jetted pocket

The pocket is made by slashing through the garment to the required length of the finished pocket. The edges are then bound and a pocket bag sewn to the wrong side of the garment.

8.1.4 Side pocket

The side pocket is set into the side seam of garment, resembling to the hidden inseam pocket. Except that the side pocket is visible on the right side. It's applied on trousers and skirts.

8.2 Collars

A collar is the strip of fabric that is meant to neaten and enhance the style of a neckline. Collars form a beautiful method of neatening a neckline. Collars can be of any shape and size.

Categories of collars:

1. Flat collars

    i   Peter pan collar
    ii Eton collar
    iii Sailor collar

3. Standing collars

    i. Mandarin collar
    ii. Polo collar
    iii. Shirt collar
    iv. convertible collar
    v. Wing collar

4. Collars cut in one with the garment

    i   Roll collar
    ii Rever front collar
    iii Shawl collar

5. Collars with revers

    i   Gents collar
    ii Reefer collar

Collars are made and sewn to the garment before seaming the side seams and setting in the sleeves. Before attaching collars shoulder seams must be stitched and neatened first. Collars can be attached in the following three ways:

1. Self-neatening collars. These are simple collars with straight neck edge, used on a shirt styled neckline.
2. Collars attached with the use of a facing, this method is suitable for straight or curved flat collars.
3. Collars attached with the use of a crossway strip, a method generally for curved flat collars.

8.2.1 Making of a straight collar

1. Place two pieces of collar with R.S. facing and matching fitting lines. Pin tack and machine stitch through the fitting lines on the three sides of the collar, start and finish the stitching at neck fitting line

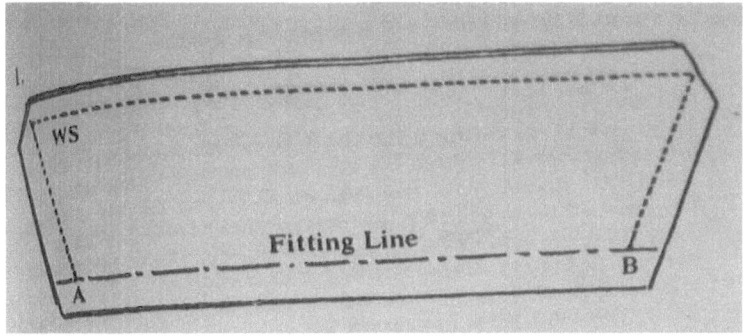

2. Trim turnings to 0.3 cm and trim off corners
3. Turn collar R.S. out and bring stitched line up on to the fold, tack around the edge to the neck fitting line only If fabric is slippery baste the two layers together

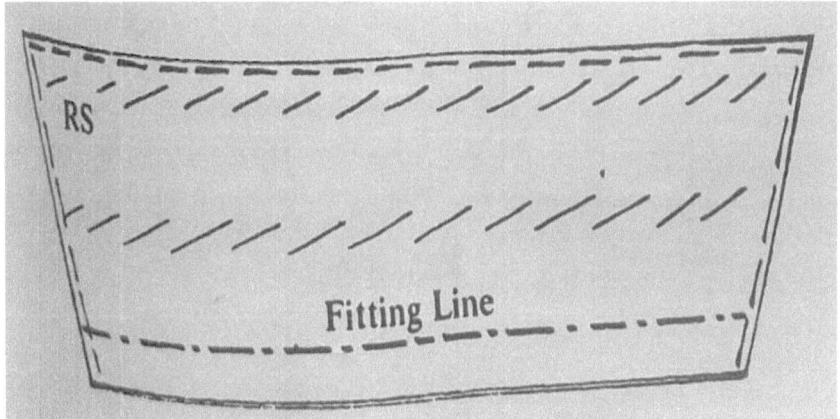

NOTE: if the outer edge of the collar is curved, snip V-shaped notches out of the turnings along the curve.

To Attach a Straight Self-neatening Collar

1. Make up collar as directed. Neaten the raw edge on the front facing.
2. Fold each front facing on to the R.S. of the garment on the fold line and match C.F lines. Pin and tack on neck fitting lines. Stitch along fitting line from the fold to C.F. line exactly A to B

on the diagram below. Fasten off ends securely, remove tacks and press.

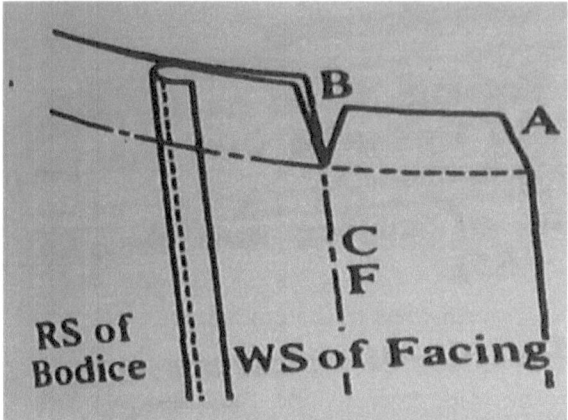

3. Snip across turnings at C.F. lines (Point B) and trim turning and corners to 0.3 cm above stitched line. Turn facings over to W.S. and ease out corners gently, bringing stitched line on the fold and press.

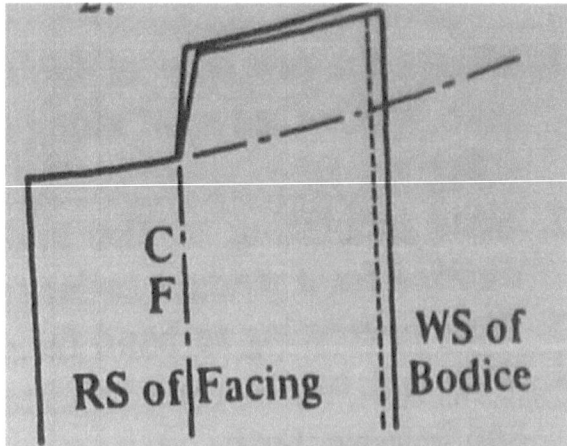

4. With upperside of collar against the W.S of bodice, place edges of collar to the C.F. lines. Fold turnings of under collar away towards worker. Matching neck fitting lines of bodice and upper collar only, pin tack, and machine stitch. If necessary ease garment on to collar.) Fasten off ends securely, remove tacks and press. Trim turnings to 0.6 cm and snip where necessary.

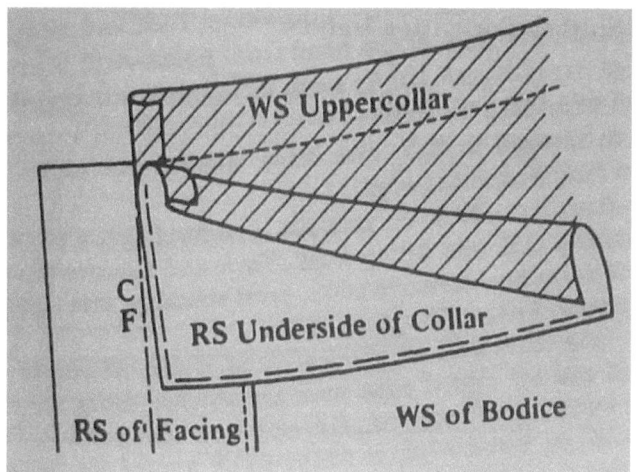

5. Turn the collar up and over turnings. Press collar up from the stitched line so that turnings are enclosed.
6. Working from the R.S. of garment, fold under the turnings of the under collar and lay the folded edge along the stitched fitting line, so that all turnings are inside, pin and tack into position. Hem the fold on to the machine stitching. Remove all tacks and press.

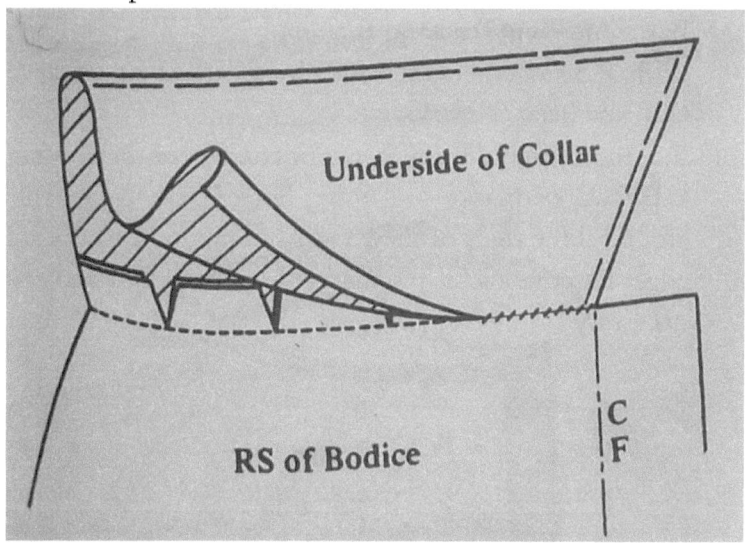

NOTE: When collar is folded in to place, the hemming is concealed.

To Attach a Shirt Collar with a Band

1. Neaten the raw edge of the front facing. Fold each front facing onto W.S. of garment along the fold line and tack into position.

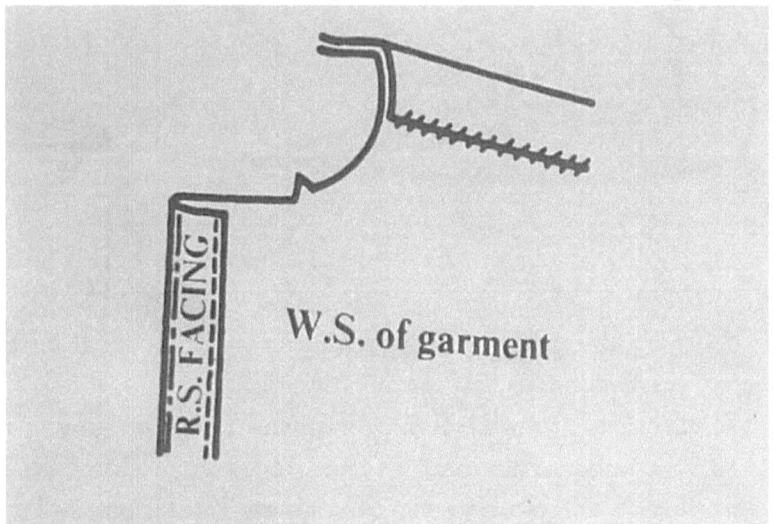

2. Baste interfacing to the underside of collar and make up as directed for a straight collar.
3. Baste interfacing to band facing.
4. With R.S. together pin band to underside of collar matching C.B. and balance marks.
5. R.S. together, pin band facing to band, over the collar matching C.B. balance marks and fitting lines. Tack and machine stitch around three sides of the band from points A to B leaving neck edge. Remove tacks, press stitching, trim seam and snip where necessary.

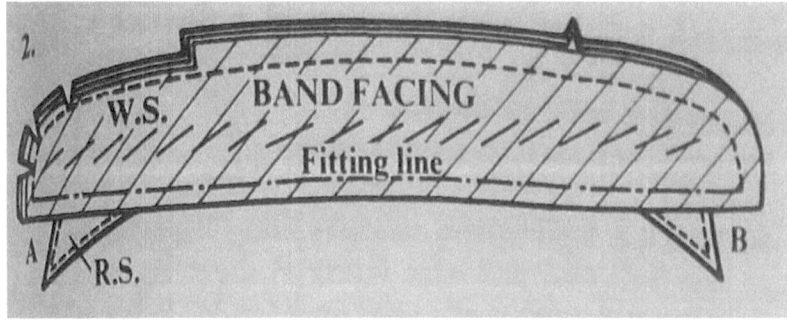

6. Turn through to R.S. and bring stitched line up on the fold and tack around the edge.
7. Working from W.S. of garment, pin band facing to neck edge matching C.B. and fitting lines. Tack and machine stitch down neck fitting line. Remove tacks, press stitching, trim turnings and snip curved edges.

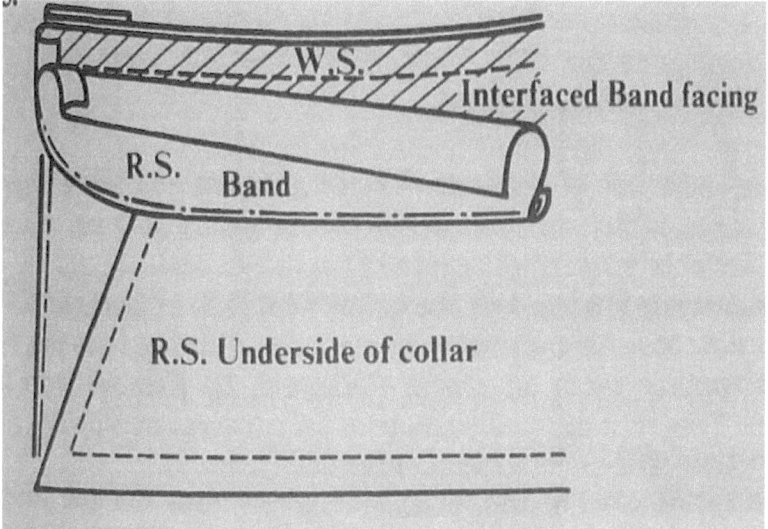

8. Working from R.S. of garment, fold under the turning on the band and lay folded edge over the stitched fitting line so that the turnings are enclosed. Pin and tack into position. Starting at C.B. on collar edge, machine stitch around the band

overlapping for ½" at join and remove all tacks and press.

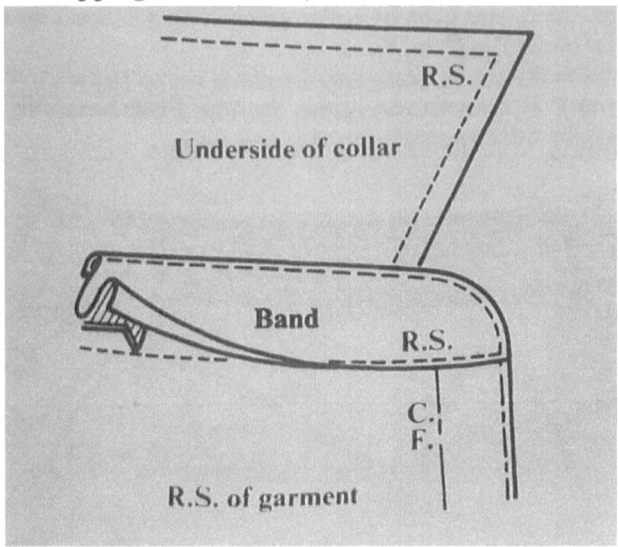

To Attach a Collar with the Use of a Shaped Facing

1. Make up the collar as directed
2. Prepare the shaped facing:
    (a) With R.S. together and matching fitting lines, join back neck and front neck and edge facing at shoulders, with a plain seam; trim turnings and press open (on full length openings a further join in the front facing is often required).
    (b) Neaten the outer edge of the facing all round by folding 0.6 cm turnings to the W.S. tack, machine stitch, close to the fold. Remove tacks and press.
3. Place under side of collar to R.S. of garment and bring edges to C.F. lines. Matching balance marks and fitting lines pin and tack the collar in position.

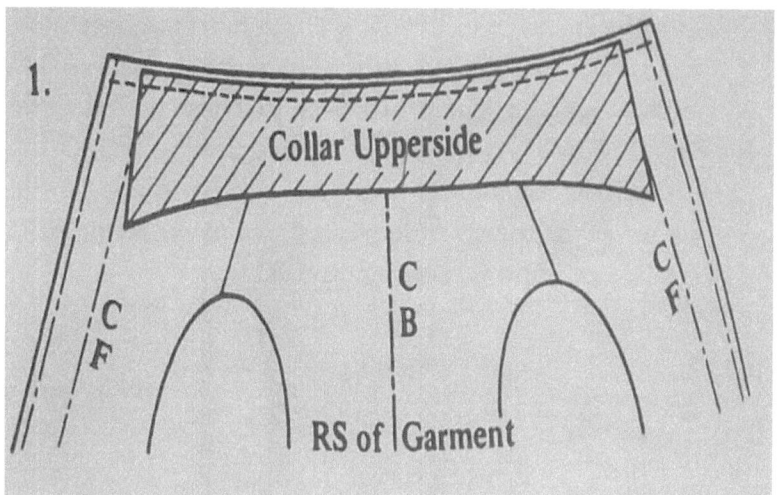

4. Place prepared facing over the collar with R.S. of facing to R.S. of garment. Matching all balance marks and fitting lines pin, tack and machine stitch all round. Remove tacks and press. Trim turnings to 0.6 cm, trim off corners and snip into seam allowance of the curved neck edge

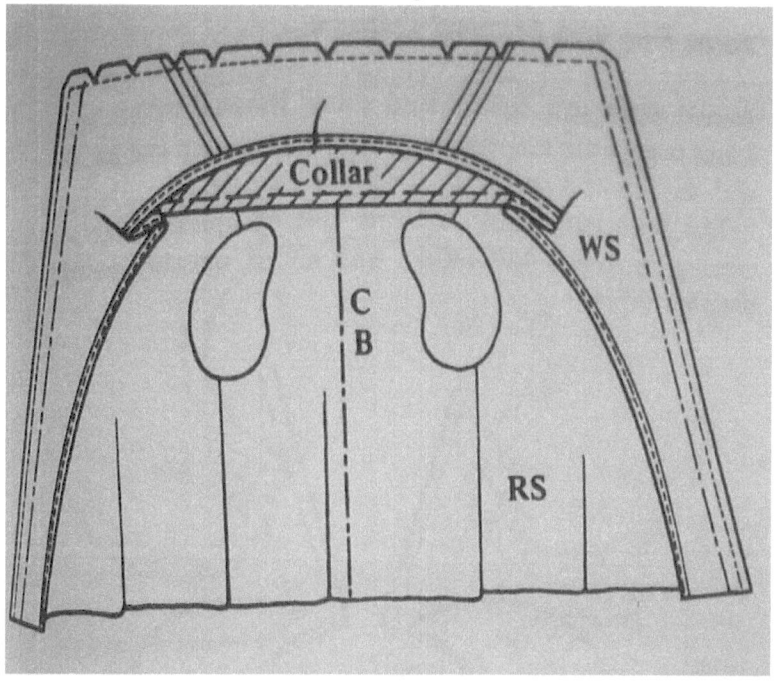

5. Fold facing over to the W.S. and gently ease out the corners. Bring stitched line on to the fold and press carefully. Tack around edge of facing and base of collar to hold in position until garment is completed.
6. Secure the facing by hemming in place on to the seam allowance of the shoulder seams. The front facing will be held in place by buttons, buttonholes and hem.

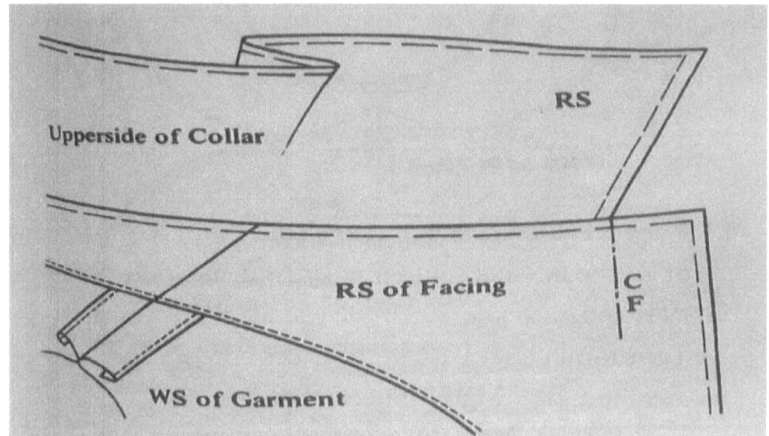

8.2.2 Peter Pan and shaped collars

Pater pan is a flat and Round shaped collar. The collar can be attached with the use of a facing or a cross way strip. It consist of two opposite collars that are made and shaped in the same way it can also be made as one complete collar with an opening at the centre back.

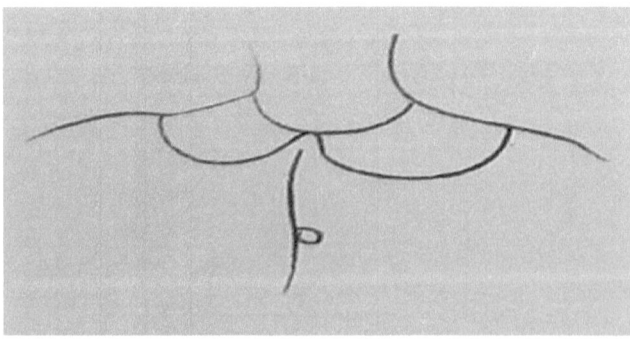

To make a Peter Pan or Shaped Collar

(a) Place the two pieces of collar R.S. together. Matching fitting lines, pin, tack and stitch outer edge of collar Remove tacks
(b) Trim turning to 0.6-0.3 cm according to fabric used. Snip notches into all curved turnings

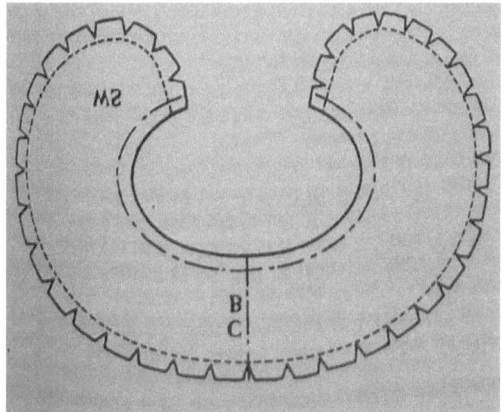

(c) Turn collar R.S. out and bring stitched line on to the fold, press carefully. Tack round the edge to keep the fold in place. If fabric is slippery baste the two layers together.

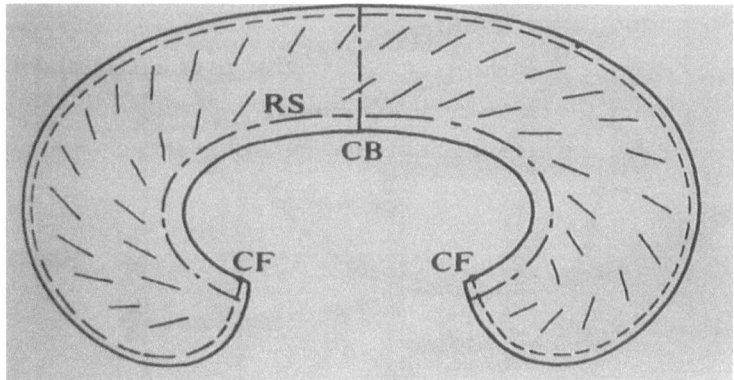

To Attach a Collar with the Use of a Crossway Strip

1. Make up collar as directed. Neaten the raw edge of centre back or centre front facing extension. Cut a crossway strip 2.5 cm wide and length of neck edge.
2. Place underside of collar to right side of garment and bring edges to meet centre front and centre back lines. If two half collars, are used, they should meet exactly on the centre front

line at neck fitting line. Match all balance marks and fitting lines, pin and tack collar in position, easing the bodice on to the collar if necessary

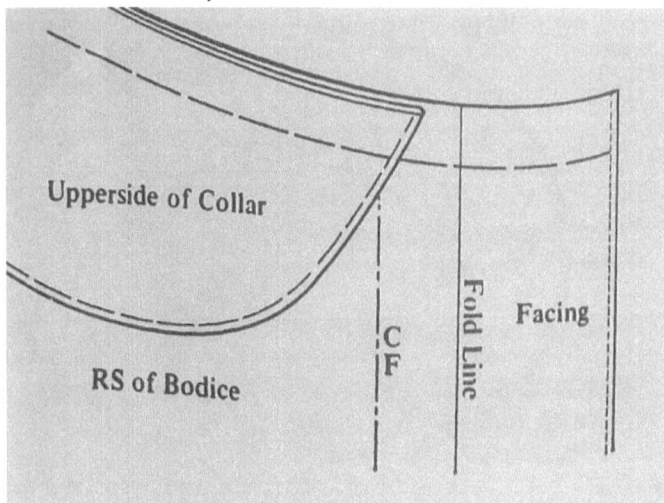

3. Fold on the extension lines and turn facing back on to right side and over the collar matching centre lines. Pin and tack into position at neck fitting line
4. Place right side of crossway strip over collar and garment, placing the raw edge 0.5 cm above the fitting line. Pin and tack so that the crossway overlaps the facing at each side by at least 1.25 cm.

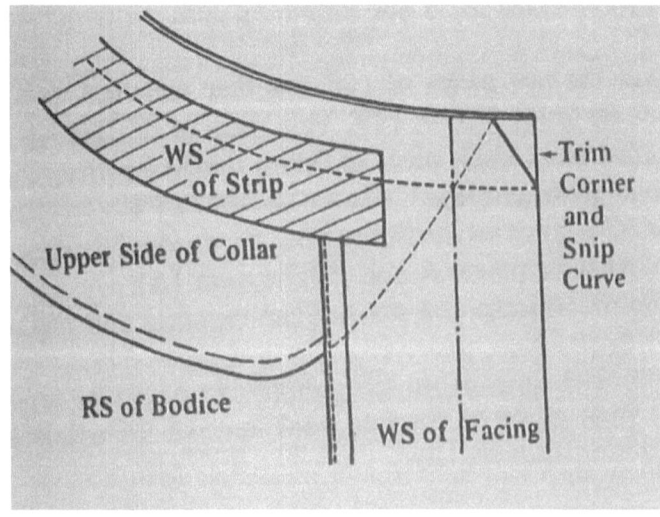

5. Machine stitch from edge of extension. Fasten off ends securely, remove tacks and press. Trim turnings to 0.6 cm, trim corners and snip into curved edges.
6. Turn strip up from stitched line and press. Turn strip and edge facing on to wrong side and gently ease out the corners. Crossway strip now forms a narrow facing. Tack against base of collar to hold edge of facing in place.
7. Turn under 0.6 cm along raw edge of the crossway strip and, allowing this fold to stretch slightly, pin and tack the crossway flat on to the garment.

Starting at point A hem the neatened edge of the facing on to the crossway, then continue to hem the edge of the crossway in place. To neaten the hem second facing to crossway, remove tacks and press.

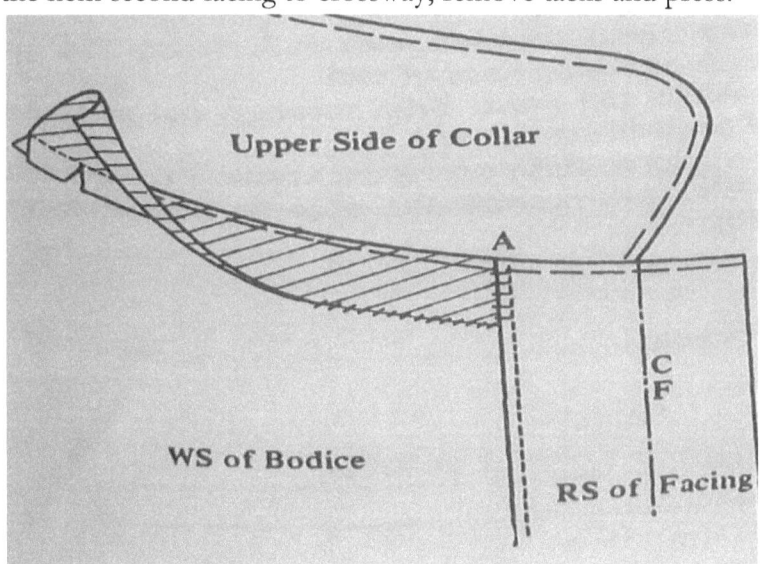

8.3 Revision Questions
1. Outline the methods of attaching pockets
2. Explain how to attach a patch pocket
3. List the methods of attaching collars
4. Define a welt pocket

5. Why are corners of the patch pocket trimmed before being sewn onto a garment?
6. Define a collar
7. List the categories of collars.
8. List the types of standing collars

Chapter 9: Sleeves and Cuffs

9.1 Sleeves

Sleeves have been used to neaten the armhole as well as changing the outline of a dress. Important sleeve outlines and designs keep changing and back over a period of time. There are two main categories of sleeves: the set in sleeve cut separately and then set into the armhole of the garment, and the sleeve cut together with part or the full bodice.

Types of sleeves:

- Cap sleeve - The sleeve is set away from the arm and can be shaped into different designs.
- Puff sleeve - Puff sleeves are made by adding extra material to the sleeve's width, the fullness is then gathered into the armhole and cuff respectively. .
- Bell sleeve - Bell sleeves fits smoothly into the arm hole and flaring out in the shape of a bell which can be of any length and flare.
- Leg-of-Mutton Sleeve - This sleeve fits into the armhole then fullness is added from the shoulder to the elbow then fitting from the elbow to the wrist.
- Raglan sleeve - The raglan sleeve is cut to include part of the neckline and armhole. The raglan sleeve can be developed for bodice, dress, blouse, jacket, coat etc.
- Kimono sleeve – the kimono sleeve is developed and cut in one with the garment, thus the sleeve is never separated from the garment.

9.1.1 Set-in Sleeves

Great care must be observed while attaching a set in sleeve to ensure a comfortable fit and a well-made look. It is essential to trying on the garment with the sleeves tacked in would be of essence to check that the sleeve fit and hang correctly.. Sleeve should be set in to the arm hole with seams neatened and wrist or lower edges completed. Collar or neck edge should be completed before setting in the sleeve. Follow pattern instructions carefully to ensure that the correct sleeve is in the correct armhole.

Making up the Sleeve

1. Run a gathering thread on the fitting line around sleeve head between the balance marks. Run a second thread 0.3 cm above the fitting line. For ease in drawing up even disposal of fullness, work this line in the opposite direction (Approx. 3.75 cm extra fabric is allowed for ease of movement.)

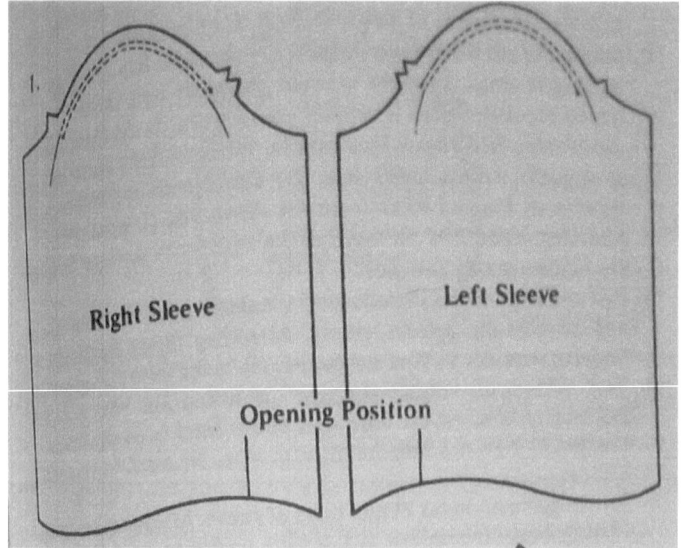

2. Make darts at elbows if necessary and press downwards
3. Matching balance marks and fitting lines, pin, tack and finish under-arm seam or panel seams.
4. Complete the lower or wrist edge.

Setting in a Sleeve

A good set-in sleeve should be set with sleeve head free of pleats or gathers that fits on the arm hole accurately. This can be achieved by making ease stitches used to ease the fabric.

1. Sew the shoulder seams and the side seams of the garment and press them open.
2. Sew the seam of sleeve and press open and turn the sleeve to the right side.

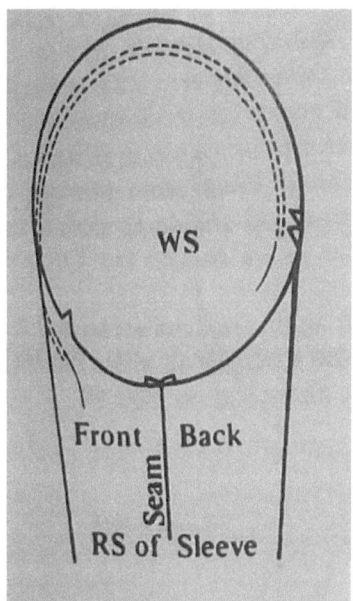

3. Work two rows of easing stitches between the balance marks 1cm from sleeve edge and the second one at 1.2 cm.
4. Insert the sleeve into the armhole right side facing make sure that the notches and underarm seams are matching.
5. Match the shoulder to the highest point of the sleeve head.
6. Gently pull the easing stitches till the sleeve fits in to the armhole and pin the sleeve in place.
7. Starting from the underarm seam machine stich the sleeve, sew straight over the shoulder and overlap the stitch on the underarm to reinforce the stitching. A seam allowance of 1.5 should be maintained.

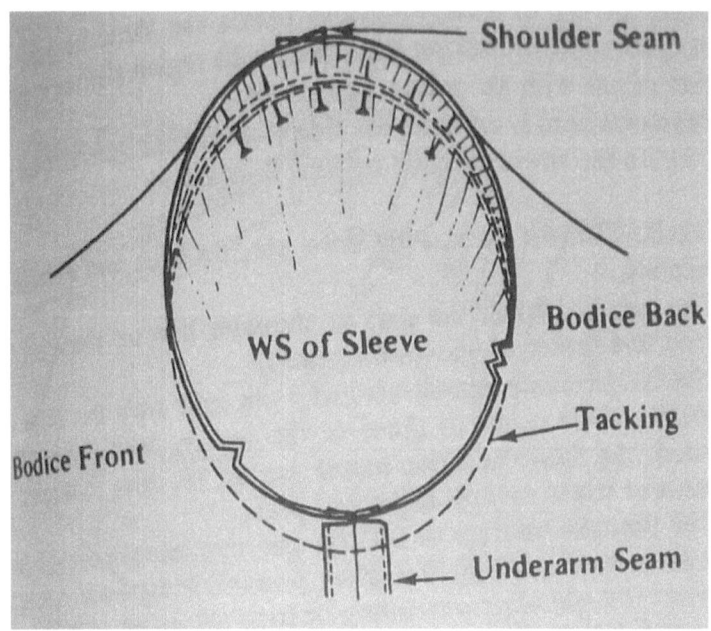

8. Neaten the raw edges of the seam by use of a zigzag stitch, binding or overlock stitch.

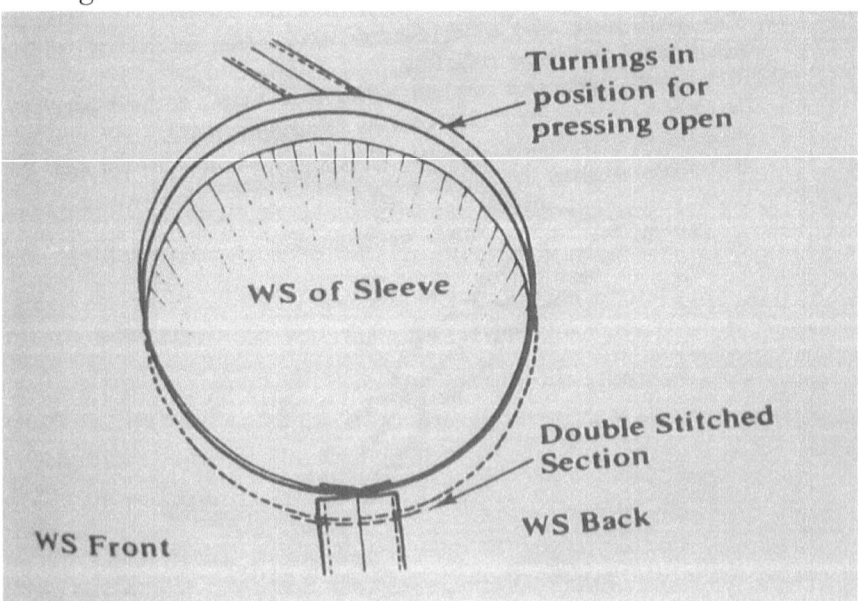

9.1.2 Raglan Sleeve

The raglan sleeve forms part of the bodice the order will be differ from that of the ordinary armhole.

The sleeve is attached to the bodice the before side seams of the garment are stitched. The sleeve is made up and finished after it has been attached to the bodice.

1. Stitch and finish darts, style lines, etc., on back and front bodice sections.
2. Pin, tack and stitch the dart at shoulder line on sleeve section. Press and neaten Matching balance marks carefully pin and tack the front and back armhole seams to those of the bodice, taking care not to stretch the fabric as these seams are on the. Machine stitch, remove tacks and press.

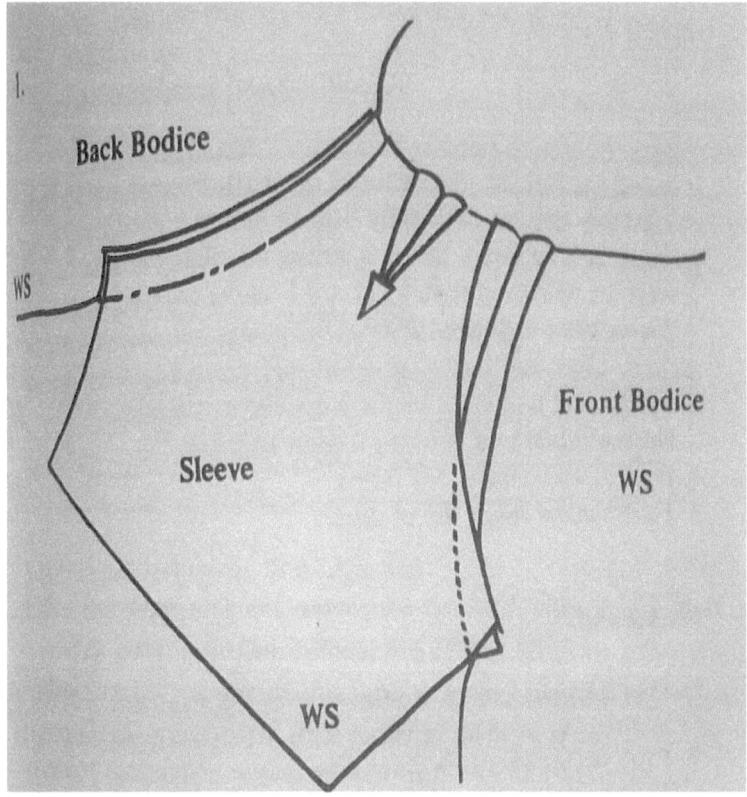

3. Trim the seam turnings to 1.25 cm and snip into the curved edges to allow turnings to lie flat when pressed. Press seam open and neaten raw edges by overcasting or loop stitch.
4. Pin and tack the under-arm and side seam of bodice matching seam and balance marks.
5. To strengthen the seams at under-arm, tack a piece of tape to the fitting line around the curve.
6. Machine stitch the seam. Remove tacks, press and neaten as the armhole seams.

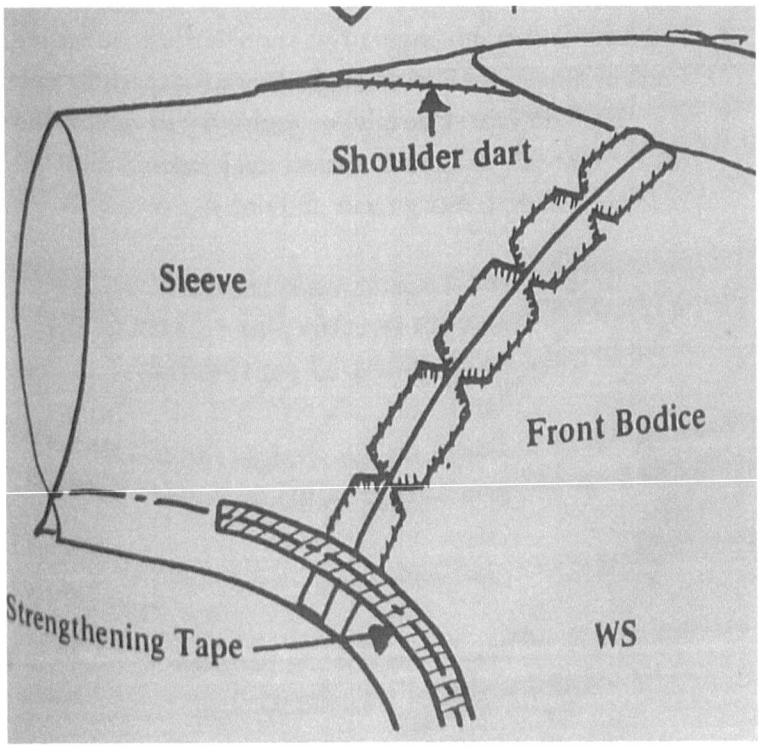

9.2 Cuffs

A cuff is a stripe of fabric sewn to the lower edge of the sleeve of a garment used to finish the raw edges of a sleeve, coat, dress or trouser leg and to protect the garment edges from fraying. Shirt cuffs normally have open edges that fastened together to fit wrist. Cuffs can also be made with elastic to allow them stretch and still fit.

9.2.1 Cuff with an overlap

A general method for a self-neatening cuff.

1. Place the two pieces of cuff together with right side facing. Pin and tack on fitting lines of cuff. This form of straight cuff can also be made from a single piece of fabric folded lengthways as for waist band.
2. Machine stitch on the fitting line, starting and finishing exactly on the wristline points A and B

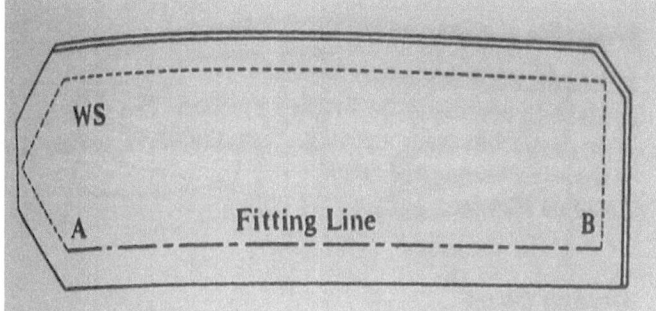

3. Remove tackings and press. Trim turnings and snip off corners.
4. Turn cuffs through R.S. easing out points on corners. Bring the seam up on to the fold, tack around the edge to keep it in position

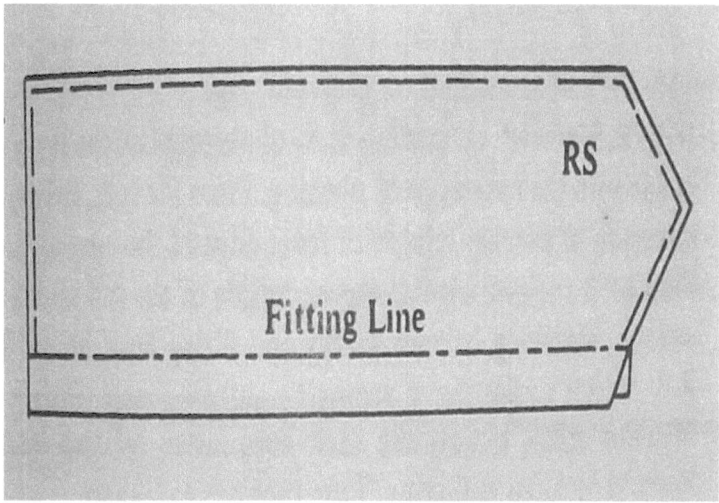

9.2.2 Open or shaped cuff

1. Place the two pieces together with right side facing. Pin, tack and machine stitch on fitting lines of cuff.
2. Remove tackings and press. Trim turnings and points and snip into curbed seam.

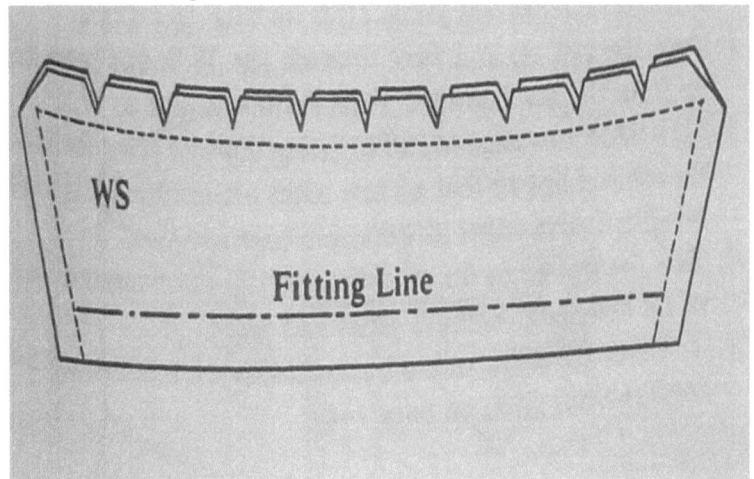

3. Turn cuff through to right side, easing out corners. Bring the seam up to the fold, tack around the edge to keep it in position.

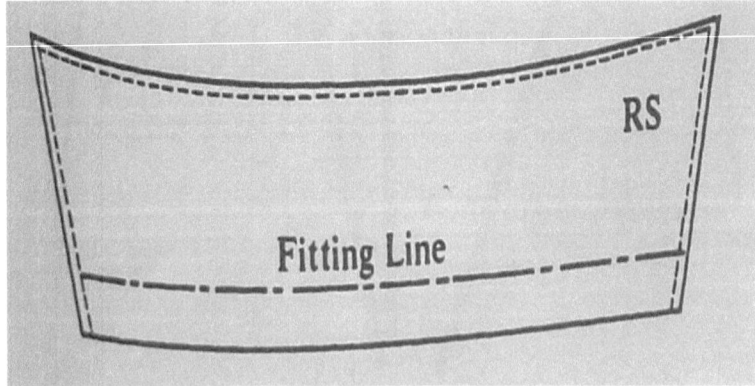

To attach a Self-neatening Cuff with Overlap

Preparation of sleeve

1. Join the sleeve seams and neaten

2. Make an opening at the marked position. This could be a faced or continuous strip opening. Use the most suitable method according to style and fabric.
3. Gather the edge.

Attaching the cuff

Prepare the cuff following the method already given.

1. With R.S. of cuff against the R.S. of sleeve place the square end in line with the back edge of opening. Place the tack, marking the extension of overlap level with front edge of the opening.
2. Fold back the raw edge of the underside of the cuff towards the worker. Matching balance marks and fitting lines, pin and tack R.S. of cuff to the sleeve arranging gathers evenly.

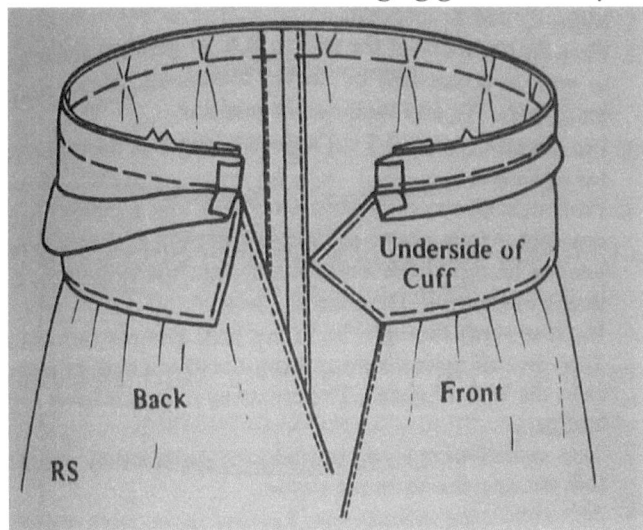

3. Machine stitch against the tack line, fasten off ends securely. Remove tackings and press. Trim turnings to 1.25 cm.
4. Turn the cuff up and over towards the W.S. and press the cuff up from the stitched line. Turn sleeve through to W.S.
5. Turn under raw edge of cuff to fitting line and bring the fold on to the stitched line so that all raw edges are enclosed. Pin and tack, bringing folded edges of extension together also.

6. Hem the fold on to the stitching and slip stitch together the edges of the extension. Remove tacks and press.
7. To fasten the cuffs attach a button and work a buttonhole

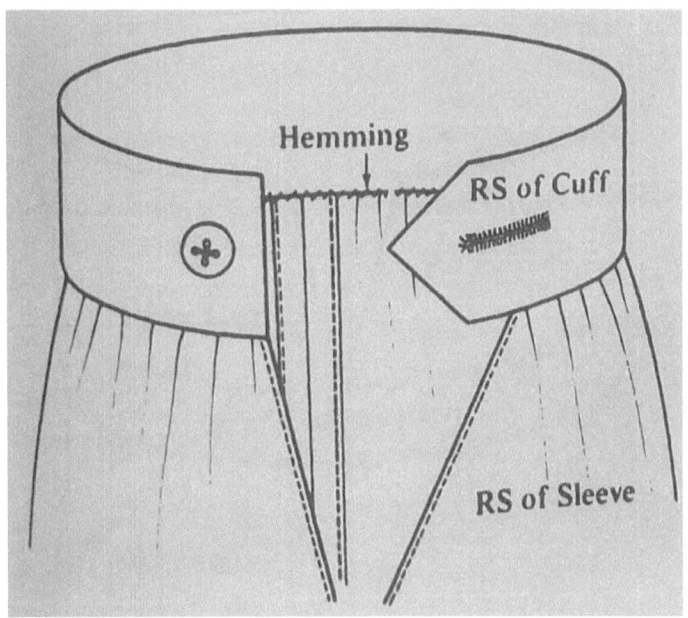

To Attach a Cuff with the Use of a Crossway Strip

1. Make up cuff as directed
2. Place the underside of the cuff to R.S. of garment and bring edges to meet at centre line of sleeve. Match all balance marks and fitting lines. Pin and tack cuff in position.
3. Cut a crossway strip 2.5 cm wide the length of the cuff + 1.256 cm for turnings.
4. Place R.S. of crossway strip over cuff and garment placing the raw edge 0.5 cm above the fitting line. Pin and tack so that the join lies on top if the sleeve seam. Join the strip on the straight thread as shown in.

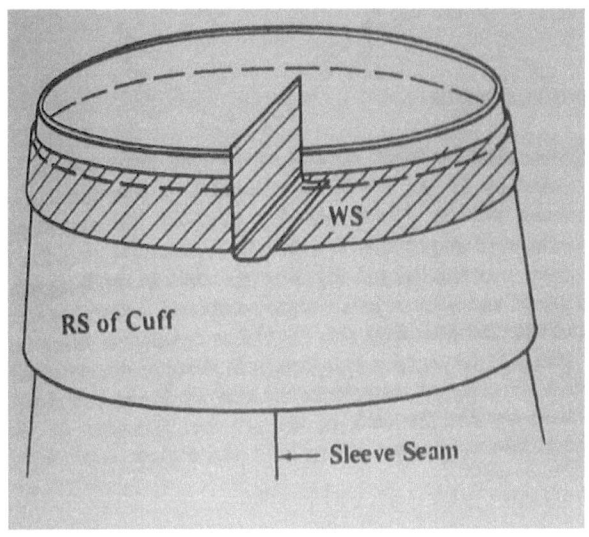

5. Machine stitch through the fitting line. Remove tack and press.
6. Turn the cuff upwards away from the sleeve and the strip down on to the W.S. of sleeve. The crossway strip now forms a narrow facing.
7. Turn under 0.6 cm along raw edge of the crossway strip, pin and tack the strip flat on to the sleeve.
8. Hem around the tacked edge. Remove tacks, press, and fold cuff back into position.

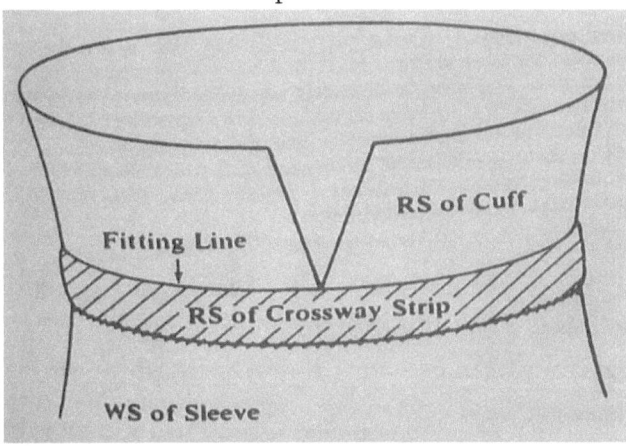

One-piece cuff

A one-piece cuff is cut out as a single piece of fabric as one and mostly only half of the cuff is interfaced.

1. Fusible interfacing is applied to the half of the cuff that will be the right side.
2. Turn seam allowance on the side that is not interfaced to the wrong side and tack in place.
3. Fold the cuff in to two right side facing.
4. stitch the short ends of the cuff
5. trimming one edge the seam and snip the corners
6. Turn the cuff to the right side and press.

Two-piece cuff

1. Apply fusible interfacing to the top cuff
2. Turn seam allowance of the lower cuff to the wrong side and tack in place.
3. Place the top and under cuff together with the right sides facing
4. Starting with one side of the cuff machine stitch the short end together. Followed by the along the lower edge.
5. layer the seam and snip the corner and turn the cuff to the right side
6. The cuff can now be stitched to the sleeve.

9.3 Revision Questions
1. List types of sleeves
2. State two major classifications of sleeves.
3. Give examples of the classifications of sleeves.
   a. Sleeve that is combined with part of the bodice.
   b. Sleeve that is combined with the entire bodice
4. Why is it necessary to run gathering threads around the sleeve head before setting it in?

Chapter 10: Waistbands and Belts, and Hems
Introduction

It is a necessity that the waist line is finished accurately in order to fit comfortably and securely. Waist bands are used on skirts, shorts and trousers. A strong interlining or petroleum is used to prevent the waistband from stretching as it's always under strain.

The skirt waist band is attached after every detail has been constructed. Pockets, pleats and seams neatened and pressed and the opening should be completed first. The hem line is measured and turned after the waistline has been finished.

The two methods of finishing a waistline

1. Use of a waistband cut from fabric same as that of the garment (self-fabric)
2. Stiff ribbon used to reinforce waistbands (Petersham) set inside top of skirt

10.1 Waistbands

It's recommended that a waistband should overlap when fastened to obtain a good finis. One would determine to either make a pointed extension of the overlap or a square extension of the underlap

The waistband should be cut on the straight grain, lengthwise parallel to the selvedge. When using by the commercial pattern cut the band as directed.

(a) Length required – which is firm waist measurement plus 2.5 cm plus 3.75 cm allowance for overlap plus 1.25 cm turnings x
(b) Twice the finished width plus 1.25 cm turnings.

The interlining also must be cut on the straight thread; use a suitable interlining, or a strong calico or bonded stiffening.

Cut the interlining –

(a) Finished length of band plus 0.6 cm turnings
(b) The exact finished width of waistband

10.1.1 Making Waistbands

1. Mark the fold line clearly, also fitting line and shape of overlap if pointed.
2. Baste the interfacing into position.

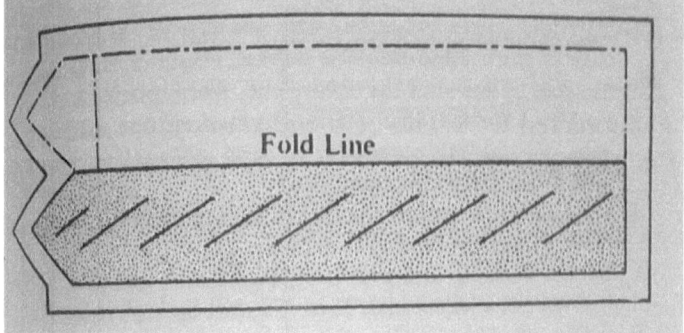

3. Fold in half lengthwise, R.S. together
4. Tack and stitch each end exactly on the fitting line, nor across turnings. Press. Trim end turnings and point closely.

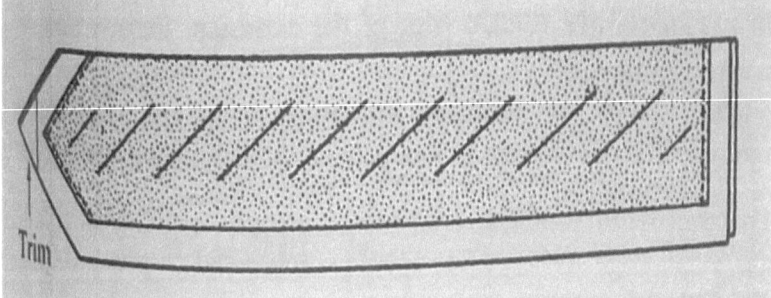

5. Turn band through to R.S. easing out point and corners.
6. The same method is used for preparing square ended band with underlap extension.

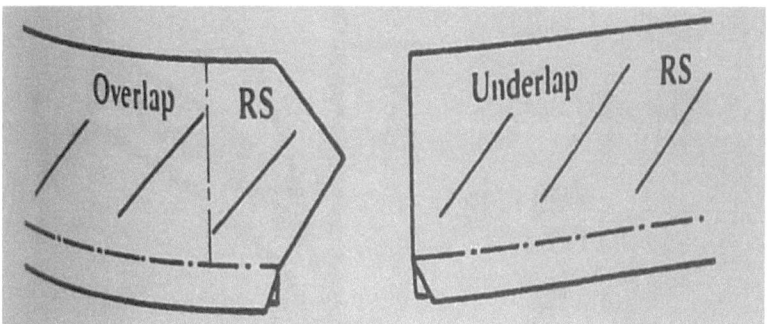

10.1.2 Setting on waistband

Pointed Extension of Overlap:

1. With the fold of the band towards the hem, place the R.S. together, the stiffened side of the band against the R.S. of the skirt.
2. Place the square end in line with the edge of the opening on the back skirt. Place the tack marked length of the extension, so that it is in line with the square end, when opening is closed.
3. Fold back the raw edge nearest the marker, then matching fitting lines, pin and tack outer edge of band to waist line, easing in the skirt if necessary
4. Machine stitch against the tacked line, remove the tackings and press. Trim skirt turning to 1.25 cm.

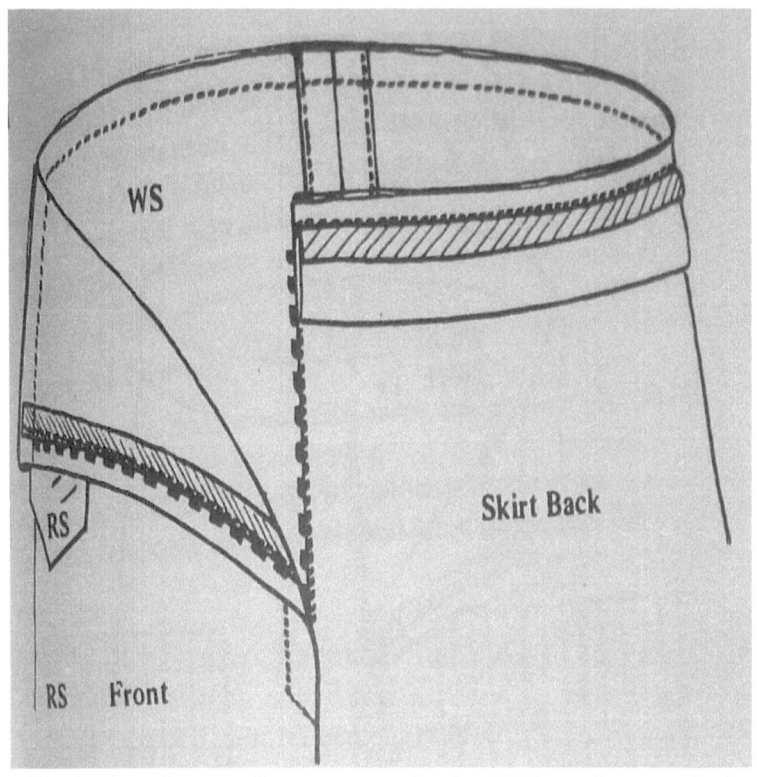

5. Lift the band up and over towards the W.S. and press the band up from the stitched line. Turn the skirt through to the W.S.
6. Fold under the raw edge of the band to fitting line and bring the fold down to machine stitching, so that all raw edges are enclosed. Pin and tack.
7. Hem the folded edge on to the machine stitching and slip stitch together the straight edge of the extension. Remove tacks and basting. Press.
8. Complete the waistband with three hooks and bars placed as shown in.

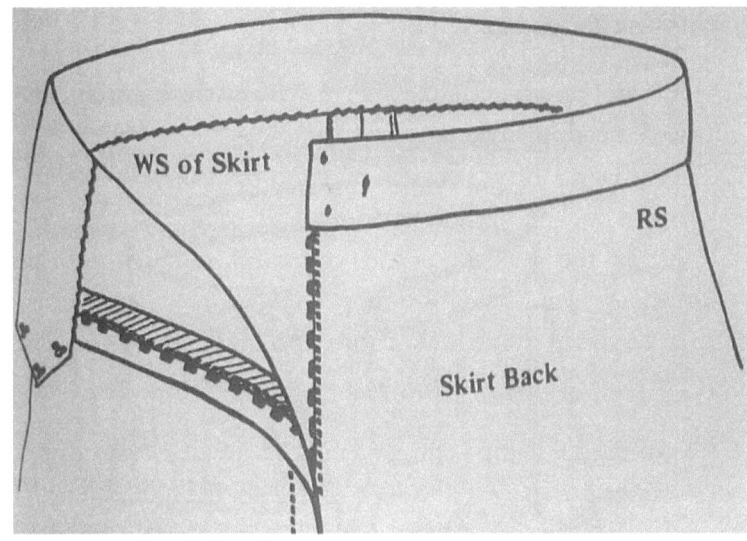
Square End with Overlap

Follow the method as given, except place the square end of the overlap level with the edge of opening on front of the skirt, therefore the underlap is extended.

NOTE: Position of hooks on square end.

Alternative Method with Top Stitched Finish

Follow the method as given except in reverse, therefore in stage 1 place the unstiffened side of the waistband against the W.S. of skirt and complete process with machine top stitching on R.S.

10.2 Petersham bands

As the petersham is not seen from R.S. it is suitable for skirts with decorative style lines such as hip yokes. It is particularly useful for finishing the waistline of bulky tweed fabrics. Petersham is available not only in plain black or white but also with spaced bones, for added support, and also with one slightly looser edge for use on hipster skirts and trousers.

Preparing the Petersham

1. Cut the length required, i.e. firm waist measurement plus 2.5 cm for ease plus 2.5 cm for neatening.
2. Turn back 1.25 cm at each end and hem the raw edges.
3. Attach two hooks and eyes.
4. Neaten the raw edges and base of fastenings with straighten binding. Hem into position.
5. Divide band into four sections and mark with a tack line.

Setting in petersham for waistline with edge to edge fastening

1. Turn garment through to W.S. and divide waist edge into four sections and tack mark.
2. With R.S. of petersham over W.S. of skirt bring the edge of petersham to waist fitting line. Place each end of petersham level of opening.
3. Matching the quarter section marks, pin and tack edge of petersham to fitting line, easing in skirt if necessary.
4. Machine stitch edge of petersham. Remove tacks and press.
5. Trim raw edges to 1 cm and neaten on R.S. with a straight binding hemmed on to the stitched line and then flat on to petersham enclosing raw edge.
6. Turn band down on to W.S. of garment and press). The hooks and bars now face in towards the skirt, in this position they are easier to fasten and prevented from catching into underwear.

10.3 Belts and belt carriers

Unstiffened belts

Cut the fabric on the straight grain with the warp threads along the length of the belt.

Length equals waist measurement plus 12.5 cm for ease, overwrap and seam allowance.

Width equals twice the finished width plus 1.25 cm turning allowance.

1. Fold strip in half lengthways with R.S. facing, pin and tack on fitting line.
2. Machine stitch on fitting lines, leaving a gap approx. 7.5 cm unstitched to allow for turning the belt through.
3. Remove tacks and press. Trim turnings, points and corners.

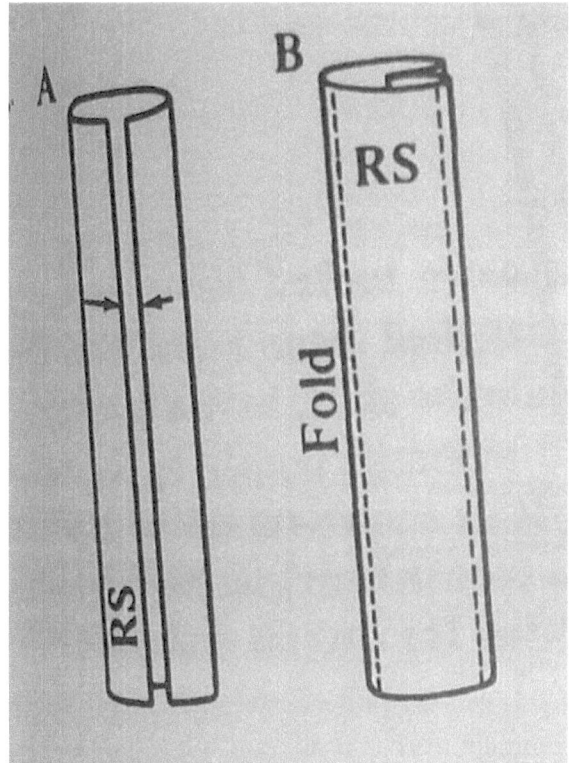

4. Turn the belt R.S. out and slipstitch the gap. Bringing the seam up on to the fold, press belt carefully. Attach required fastenings.

NOTE: Long tie belts are made in the same way but cut to the required length.

Stiffened belts

1. Prepared and make up the belt as above but leaving the buckle and open for turning through.

2. Cut the bonded belt stiffening the exact width of finished belt but 2.5 cm shorter in length to allow for buckle.
3. Tack stiffening in position on W.S. machine stitch around edge of belt except buckle end, working from R.S. to achieve a good line. Remove tacks and press.
4. Complete with buckle and metal eyelets.

Belt Carriers

Worked carrier

Work a straight loop stitched bar the width of belt plus 0.6 cm, sew on to the side seam above and below waist seam, follow method of working given for a buttonhole loop

Fabric carrier

Cut fabric on straight thread.

Length equals width of belt plus 0.6 cm plus 2 cm turning allowance.

Width equals 2.5 cm.

1. Fold the two long edges to the centre, W.S. facing Fold in half again lengthways and press.
2. Either slipstitch the folded edge or machine stitch both edges place the carrier centrally over the side seam and across the waist seam. Tack in position allowing slight ease on the carrier.
3. Sew securely in place with hemming or backstitch.

10.4 Hems

All processes must be completed before the hem is measured and turned. The garment should hang for two days to allow the fabric to 'drop', i.e. to stretch with kits own weight, before the hemline is marked, thus ensuring that it remains even after completion. The amount of 'drop' depends on the looseness of the fabric weave and the amount of 'flare' in the skirt. The drop of a circular skirt can be considerable as the true cross grain is involved.

The finished length of the skirt is influenced by current fashion and is either a particular height or depth from the knee or a specific length from the ground.

The depth of the hem is generally from 5 to 7.5 cm: on children's wear a 10 cm hem can be made to allow for letting down with growth; on thin fabrics a double hem can be turned. The method of holding the hem in position varies according to the fabric and style and the most suitable finish should be selected. The stitching of a hem should be invisible on R.S.

To mark the hem

The assistance of a second person is essential as the hemline is measured up from the floor with the help of a hem marker or yardstick held in a vertical position. The wearer should put on the garment, together with the shoes to be worn, and stand on a firm table with the weight placed evenly between both feet.

1. Using a hem marker or yardstick, mark the hemline with chalk or pins, marking not more than 10 cm apart. The wearer should check in a mirror that the desired length has been marked.
2. Take off the garment and re-mark the hem with a tack-line.

Preparation of the hem

1. Lay the garment flat on the table with W.S. outwards. Turn up the hem from the tack line, match seam lines and pin vertically at intervals. Tack through the fold 0.6 cm up from the edge.
2. With a marker card or tape measure mark the depth of the finished hem with a horizontal line of pins or with tacking taken through the single fabric only.
3. Trim away surplus fabric, leaving a 0.6 cm turning allowance beyond the marked line. Complete the hem as is most suitable.

10.4.1 Straight Hemlines: finishes

First match and pin seam lines, then the sections between seams.

Slip hemming

Slip hemming is for general use on lightweight fabrics.

To hem the garment you turn the raw edge under the marked line, then tack in position.

Slip the hem, the fold picking up one thread of R.S. fabric. Remove the tacks and press carefully from R.S. in each case.

Straight Binding

Edge Stitched Hem
1. Turn the raw edge under to the tacked line, tack and edge stitch the fold.
2. Pin and tack the stitched line into position.
3. Slip hem the fold. Remove tacks and press.

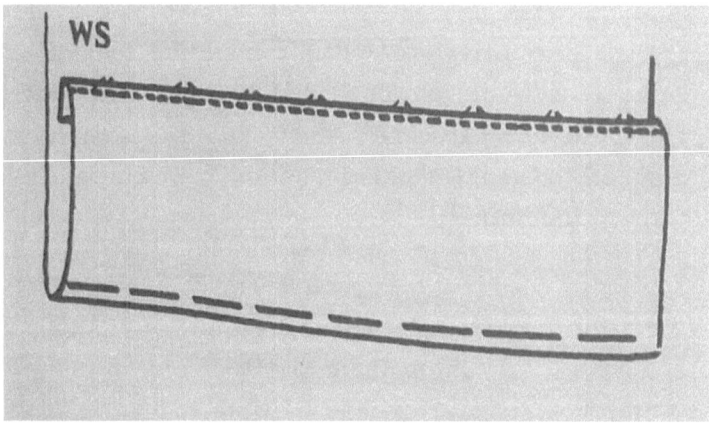

Bound Hem
Suitable for thick fabrics that fray easily. Using silk or fine, bind the raw edge of hem, following the method for binding and complete as for an edge stitched hem.

Herringbone Edge
Suitable for thick fabrics.

1. Trim turnings to marked line, i.e. the width of hem. Pin and tack into position.
2. Work herringbone stitch, picking up one thread on the R.S. of fabric and stitch normally on the hem turning.
3. Remove tacks and press.

10.4.2 Circular Hemlines: finishes

The hem is usually 0.6 cm deep as otherwise there would be too much fullness to disperse owing to the width of the flare.

1. Measure 0.6 cm down from the marked hemline and turn hem to W.S. on this line: tack
2. Edge stitch along the fold and trim away the surplus fabric: then press.

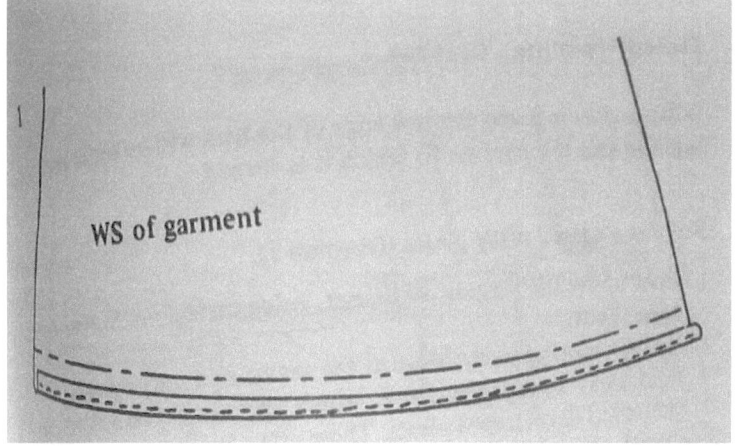

3. Turn up the hem on to W.S. on the marked line and tack.
4. Either edge stitch the fold, or slip hem stitched fold into place.

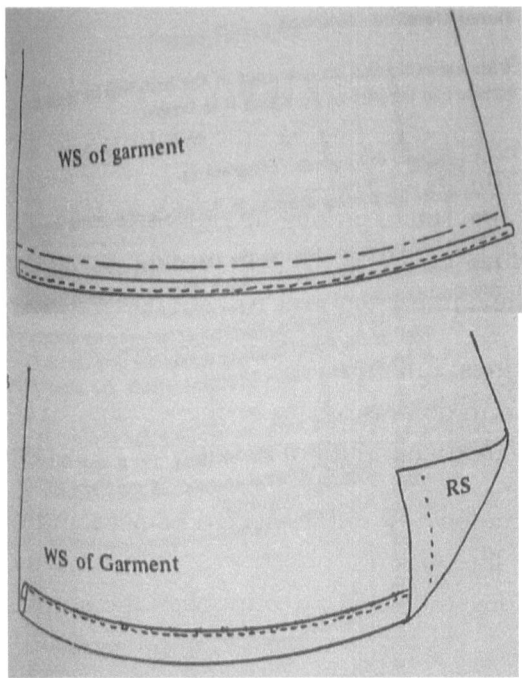

10.4.3 Flared Hemline: finishes

When a skirt is flared the raw edge of the hem will be wider than the hemline and the part on to which it is turned.

10.5 Revision Questions

1. Outline the methods of finishing a waist edge
2. Explain how to cut a skirt waistband.
3. Highlight the two types of waistband overlap
4. Why is it necessary to cut a waistband on a straight thread grain?
5. What is the reason for having children's clothes with large hem allowance that those for adults.
6. Give the types of hems

Chapter 11: Decorative Finishes

Decorative finishes are used to enhance a garments appearance. One must determine a method of decoration that is suitable for both the fabric and the style and purpose of the garment. It's essential that the method of decorative finish used on a garment must laundered is to be laundered.

11.1 Shell edging

Shell edging are decorative hems or tucks on underwear, blouses and young children's wear that is made of soft or fine fabrics that do not fray badly. Use a firm silk or cotton thread without fluff to give a crisp appearance.

Turn a narrow hem, not more than 0.6 cm on to the W.S., and tack.

Work from the right to left on the W.S., fastening on with a double back stitch on to the hem. Pick up three running stitches in the hem without going through to the R.S.

Taking the needle over the fold, pass from the R.S. through to the W.S. at base of hem, work a second oversewing stitch in the same place and pull up tightly to pinch the hem. Take three running stitches into the hem only and repeat to finish hem.

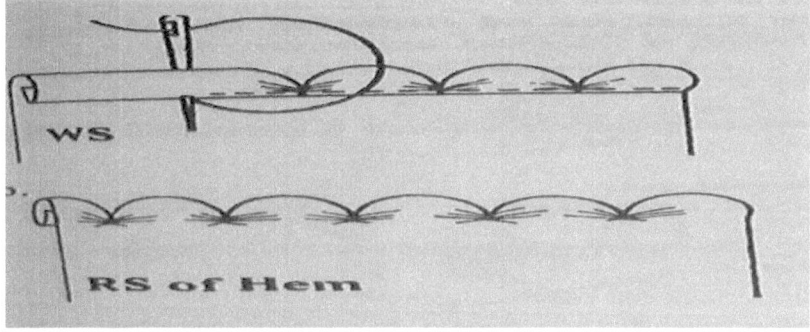

NOTE: Space between the double tight stitches should not be more than 1 cm to achieve the correct pinched effect.

Alternative method: as used for shell tucks.

11.2 Pin stitch

A decorative stitch for use on lingerie, children's wear and fine blouses.

The fabric must be fine and of a resilient weave to allow the stitches to be drawn up tightly to make a series of definite small holes.

It is worked on the R.S. to hold a folded edge on the R.S.

Uses:

1. To hold hems of any width which have either a straight or shaped inner edge
2. To stitch overlaid seams
3. To apply lace edgings or insertions
4. To apply decorative sections of applique

NOTE: Use a firm, fine, silk or cotton thread and a large needle, average size 6, to make the holes.

To pin stitch a hem

1. Turn the hem up on to R.S. and tack in position. The first fold should be evenly trimmed to not more than 0.3 cm.
2. With fold towards the worker, stitch from right to left, fastening thread with a double stitch in the fold.
3. Take a small straight stitch into the single fabric below the fold.

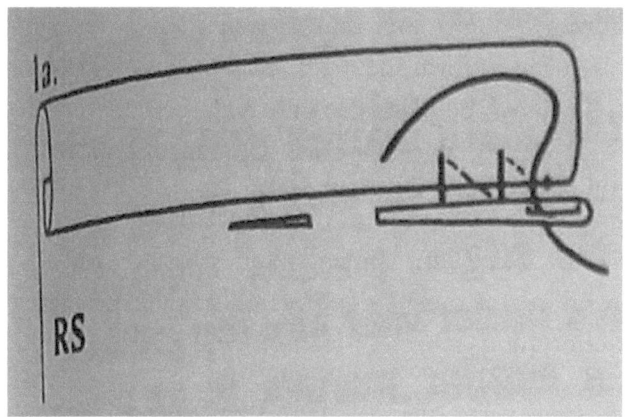

4. Repeat this stitch using precisely the same holes and draw up tightly.

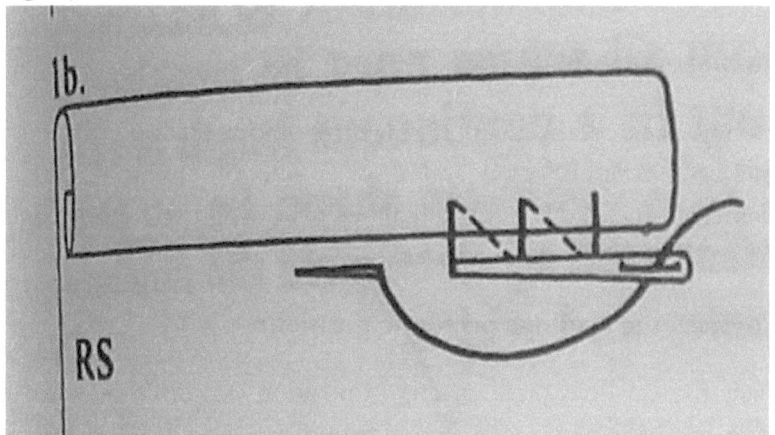

5. Place needle again into first hole and bring it out into the hem directly above the second hole, and in a square position.

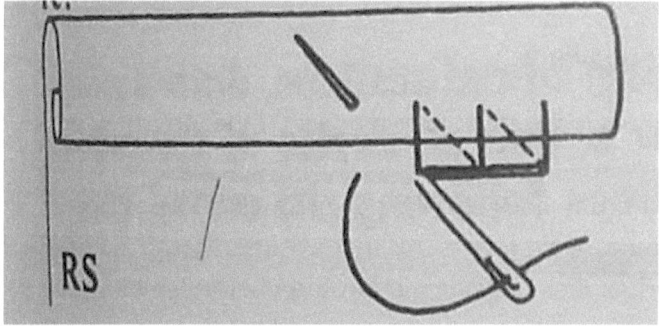

11.3 Faced scallops

Faced scallops are used to form the main feature of the design and style of the garment. They can be placed to give a shaped edge to collars or necklines, cuffs or sleeve edges, pockets, openings, and hemlines on women's and children's underwear and top garments.

Faced scallops are worked in a similar manner to an ordinary shaped facing; therefore, all seams must be completed so that the facing can be carried out in a continuous length or circle.

In order that all scallops are of a complete shape, i.e. no partial scallop, the edge must first be measured to determine the number of scallops and the necessary diameter and width.

To make a template

Having decided the diameter (width) of the scallops, draw a straight line on a strip of card and draw a line of circles of the required diameter. Accurately cut the scallops following the curved edge on one side of central line (shaped area).

To mark the scallops on to the garment

Place the card template exactly in position on the W.S. with the outer edge of the curves against the fitting line and hold firmly in place. Using a sharp piece of tailor's chalk, carefully draw around the edge of the scallops.

To face the scallops

1. Join facing as necessary and neaten the inner raw edge.
2. Place R.S. of facing R.S. of garment, matching balance marks and fitting lines exactly: pin and tack into position. Using small stitches tack round the marked line. If the facing is slippery, upright tack the facing in position before tackling the scallops.
3. Machine the curved tack line, making the points sharp between the scallops. Remove tacks and press.

4. Trim turning 0.6 cm and snip into the points almost to the stitching; snip diagonally against the curves.
5. Turn the facing to the W.S., easing out the scallops and bring the stitched line up on to the fold; tack into position. Press lightly under a cloth.
6. Hem the edge of the facing to the turning of the seam.

11.4 Faggoting

Faggoting is a way of linking together two pieces of fabric with a decorative stitch to give an open finish. It can be used to attach lace or frills to a hem, to make a decorative seam or, with the use of a Rouleau, to form an open work edging or decorative insertion.

The illustrations show the use of a Rouleau but the method is the same for applying lace, frills or for making a seam.

11.4.1 Rouleau

This is a length of crossway fabric sewn to form a long tube which can then be used for faggoting as shown, for the insertion of patterns or for buttonhole loops. As it is made from crossway fabric, it is very flexible and adapts readily to a curved position.

To prepare a rouleau
1. Cut crossway strip 2.5 cm wide and join if necessary for the required length
2. Fold in half lengthways with R.S. facing. With a loose tension, machine stitch through the centre. The loose tension makes the Rouleau more flexible.
3. Trim one end to a sharp point, thread a bodkin with a short double thread looped through the eye and sew securely to the end of the tube.
4. Push the bodkin into the tube and gently ease the rouleau through the R.S.

Prepare the garment

Neaten the raw edge by the most suitable method, either by a self-binding or a false bind equal in width to the Rouleau or a shaped facing or on firm fabrics by turning a narrow fold to the W.S. and working a running stitch along the edge; turn a second fold 0.6 cm wide and press into position.

Preparation and method of faggoting
1. On a narrow length of strong paper draw two parallel line.

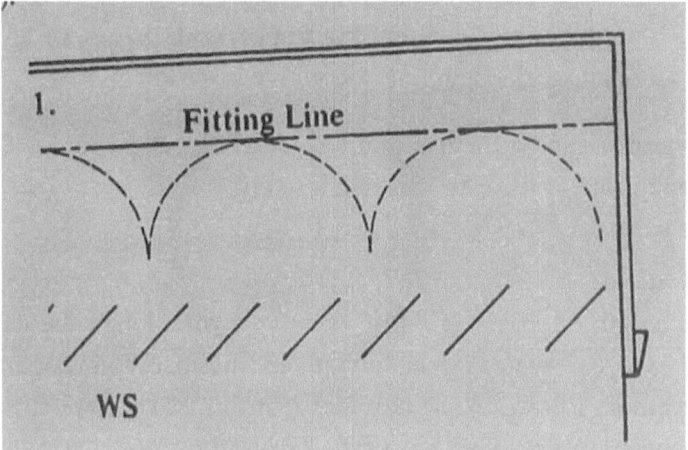

2. Tack the edge of the garment against one line and tack the stitched edge of the rouleau against the second line.

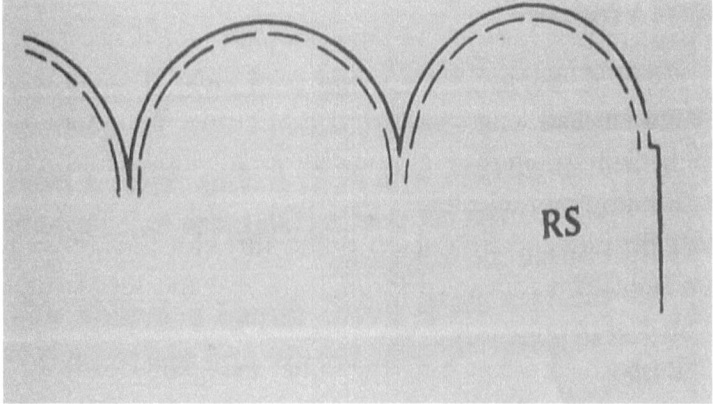

NOTE: The distance between the lines can vary according to the fabric and the choice of stiches.

3. Using a firm silk or cotton thread without fluff, fasten on into the fold of the rouleau and work stitching.. Maintain an even tension for a good finish.

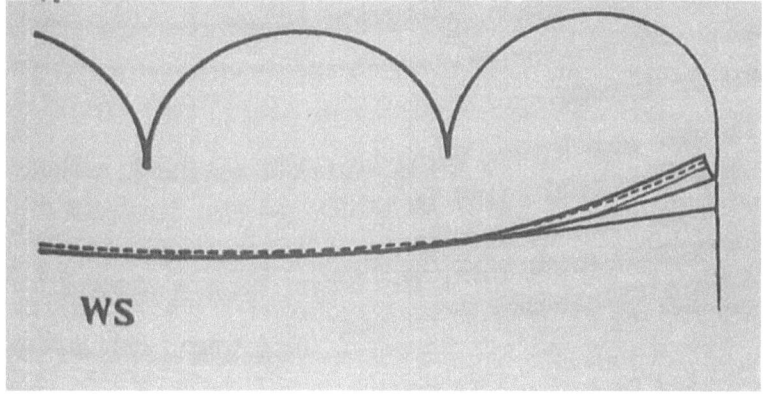

11.4.2 Frills

Frills are applied either for decoration or to lengthen a garment. The amount of fabric depends upon the desired effect.

Slight fruitless: the require length plus half as much again.

Average: twice the required length.

Very full frill: three times the required length (of fine fabric only)

To prepare a frill
1. With the warp threads running the length of the frill cut the length or length required plus turnings by the finished width plus turnings.
2. Join the lengths with a plain seam.
3. Complete the hem of the frill.
4. Work a line of gathering threads either side of the stitching line.

Lace frills: an allowance as for fabric frills. A draw thread for thread for gathering is usually to be found on the straight edge. Lace can be inset by any of the given methods.

Ready-made frilling: purchase the required length plus turnings only and inset by any of the following methods.

Methods of setting in frills

Overlaid seam Draw up the frill to the required length and make an overlaid seam with the garment section overlaid to the frill.

With a decorative band : Draw up the frill and join the garment with a plain seam made with the W.S. facing so that turnings are on the R.S. Press the garment up from the stitching against the turnings. Hem the decorative band to the frill immediately below the stitched line on R.S. and sew the second side flat on to the garment with hemming or a decorative stitch, thus enclosing the raw edges.

Inserting a frill into a panel line: Draw up the frill and tack into position on the fitting line on the R.S. of the under section. Prepare the upper section and make an overlaid seam; thus enclosing the edge of the frill.

Setting in a frill with a shaped facing: Join facing and neaten the outer edge. Draw up the frill and tack into position on the fitting line on R.S. of garment. Place the R.S. of facing over the frill and matching fitting lines, tack and stitch. Complete as for a shaped facing.

11.4.3 Lace

For joining lace use the finest thread available that is suitable for the colour and texture of the lace. For applying lace use a firm silk or cotton thread that does not fluff but which handles and looks well.

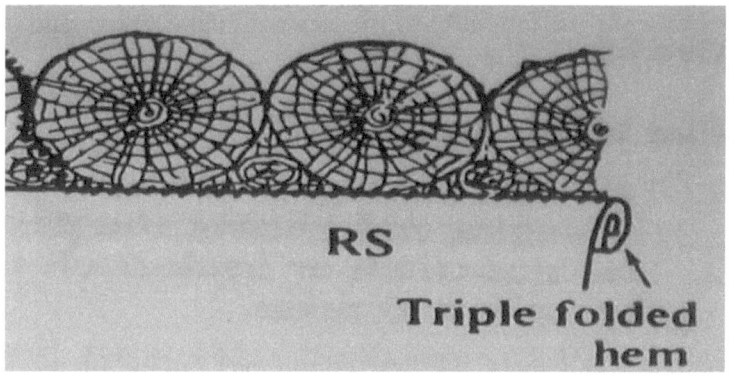

Narrow lace edging

To join

1. Match the patterns by overlaying the two layers exactly; the effect will otherwise be blurred
2. Closely oversew around the corded edge of the design to completely join the lace.
3. Trim away the raw edges close to the stitching on both R.S. or W.S.

To shape

If less than 1.25 cm wide, ease the lace into position: if more than 1.25 cm wide used without gathers, it is necessary to make small darts at corner point.

1. For the central angle of brassiere top slip sew a small dart with its point going into the angle.
2. For the shoulder strap point reverse the placing of the point.
3. On the W.S. flatten the dart and hem on to the back of the lace.

Alternate methods of attachment

1. Oversewing the edge closely to the garment and trimming away the surplus fabric: this is not suitable for gathered lace.
2. Overlay the lace on to a narrow folded hem and machine stitch.

3. Overlay the fabric on to the lace and make an overlaid seam.
4. Oversew the lace to the edge of a triple-folded hem.

Wide lace, shaped insertion and piece lace

With repeat of the design may be too large to allow for the pattern to meet exactly: the same method is sued to both join and shape the lace.

1. Mark the fitting lines on the lace and overlap the sections so that the fitting lines match
2. Select and mark the best continuous outline of the lace that passes across the fitting line.
3. Oversew very closely on marked line and trim off surplus lace.

To attach

Either over-sew (satin stitch) or pinstitch the edge of the lace on to the garment, trimming away any surplus fabric on piece lace do not make a hem but utilize the finished edge of the lace.

11.4.4 Braid

Flat braid

1. Tack the braid flat into position, making small mitred seams at the angles and joining the ends if necessary]
2. Machine stitch or backstitch or backstitch the braid along each edge. Remove the tacks and press.

Military braid and braid used as binding

On single fabric trim off allowance for turnings.

By hand: guide the braid on equally over the edge and tack through the layers. Stab stitch along the edge of the braid from the R.S. through to W.S.

By machine: open out the braid and tack into position on R.S. with ½" width of braid projecting from the edge. Machine stitch along edge of the braid. Turn the projected edge over on to the W.S. and hem on to the machine stitching.

Corners: wide corners can be negotiated by stretching the middle of the braid and easing the edges.

11.5 Revision Questions
1. List the factors to put in mind while choosing decorations for a garment
2. List the decorative finishes
3. Explain faggoting as a method of decorative finish
4. What amount of fabric is required for the frills with the following fullness
    i   Slight fullness
    ii Average
    iii Very full frill

Supplementary Content

1. Care and maintenance of clothes

To maintain the good appearance and obtain the most wear out of a garment, it is essential to take good day-to-day care of the garment and to repair at the earliest signs of wear.

General Rules

1. After use, brush outer garments and place on a hanger so that the creases drop out naturally. In the case of jumpers and knitted dresses, fold neatly and store in a drawer. Allow clothes to air overnight before storing.
2. Remove all brooches or pins before putting the garment away.
3. Try not to wear the same dress or skirt on consecutive days so that the fibres regain their original shape.
4. Check regularly that no seam is coming unstitched or that a button is loose. A button lost often means that a complete set has to be bought as it is difficult, after a period, to match up.
5. Do not put perfume on to a garment as it may stain the fabric: it will also go stale easily.
6. Check for moth. Do not store clothes that are damp or in a damp place.
7. Remove spots as soon as possible.
8. Read the maker's label for special washing or cleaning instructions before laundering.
9. Never let the garment become too soiled before washing or cleaning as heavily laundering damages the fibres and if dry cleaned the ingrained dirt will remain.

Hedge Tear Darn

Whenever possible the darning should be worked with threads of the same fabric so that the repair is less conspicuous.

1. Working on the W.S draw the edges of the tear gently together with fishbone stitch, using a very fine silk or cotton thread.

2. Mark the 'L' shape of the darn with a tack line.
3. Using the darning thread, work one side of the darn as far as the angle.
4. Work the second side of the darn completely.
5. Complete the first side. The weakest point pf the tear is then supported by double darning.

Patching

The patch should be of the same fabric as the garment or as near as possible in texture and colour.

New fabric must be washed before being used to repair a worn garment. The patch should be rectangular and large enough to cover any fabric worn around the hole. It must be cut from the same way of fabric as the part of the garment which is to be repaired, and in matching pattern lines check that the straight threads are parallel before sewing.

Patch into a seam

This method is used when the hole or worn area extends to a seam line.

1. Unpick the seam in the area affected. Where necessary remove a section of the sleeve also so that the patch may set into the seam. Outline the area to be patched with a tack line, including all the worn area.
2. Cut out the patch. On the W.S. fold over 1.25 cm turning on three sides and tack, leaving the side seam edge free.
3. Place the patch R.S. uppermost on to the R.S. of the garment over the hole. Pin and tack into position. Tack through to mark the fitting line. Trim any surplus fabric which overlaps the garment edge.
4. Oversew into position around the three tacked sides. Remove tackings and press.

5. Turn to W.S. and make two diagonal creases across the hole, from corner to corner of sewing. Place a pin 1.25 cm in from the corners on each crease. From the hole, cut up on the creases to the pins, then cut across the straight grain of fabric from pin to pin.
6. Neaten the raw edges with loop stitch.
7. Remake the unpicked section of the garment.

2. Costing (material requirements and labour charges)

Garment costing is a very important factor when it comes to garment making.

Material

Fabric is one of the major costing determinants. Material cost is determined by the quality of fabric where high quality fabric will be sold at a higher price.

A major question that arises in one's mind at the time of purchasing fabric for any garment is how much to buy? It is a very important question and to be able to give an objective reply, it requires a person to be an expert in pattern development and an expert in making an economical layout

Amount of fabric required will depend on garment design. For example a pleated or gathered garment will require more material thus will determine how much money will be spent on materials.

Pattern laying should be well planned to minimize fabric waste.

Lining

Lined garment will also have more spend on buying the lining, through cheaper than fabric. This would become the second factor in determining the cost of a garment.

Trimmings

These are all the other small items that facilitate construction of a garment including fasteners and those used for decoration.

Trimmings include

- Thread
- Buttons
- Zip
- Lace
- Ribbon
- Elastic
- Interfacing
- Shoulder pads
- Waist bands
- Hooks

Trimmings are mostly cheaper to buy but very essential for garment construction.

Labour

Labour cost may not be considered when making your own garment since in most cases stitching seen as a way of spending leisure time just like reading novels and playing.

But for commercial purposes labour becomes a very important aspect. Labour charged should be able to compensate the time and skills spend to make a garment.

Thus a simple garment that would take short time to make would therefore be charged less labour.

While a complicated design that involves a lot of time and technique high labour is charged.

Ideally, labour charged is 30% of the total material cost (fabric + lining).

Other expenses

These include

- Rent in case of a rented shop
- Electricity – used for lighting and ironing
- Water
- Fee for business permit

These expenses should be factored in at a particular percentage. Garment cost would be calculated as material cost + labour cost + trimmings + other expenses.

3. Computer-generated design and pattern making

When Computer-Aided Design (CAD) was first introduced in to clothing industry, it was really costly and could only be purchased by big companies. The whole procedure from designing to manufacturing is normally an investment taken by a manufacturer who makes a certain types of clothes in large quantities. However, small factories can gain from particular components of CAD. Application of CAD can be more profitable to a company when effectively executed in its manufacturing processes; which can be identified under these areas: product design and merchandising, production information processing and management, pattern drafting and alteration, pattern sizing and made from direct measurements, pattern grading, and lay planning and marker generation.

Chapter 12: Garment Making Practical

12.1 Skirt

12.1.1 Straight skirt

Front skirt   Back skirt

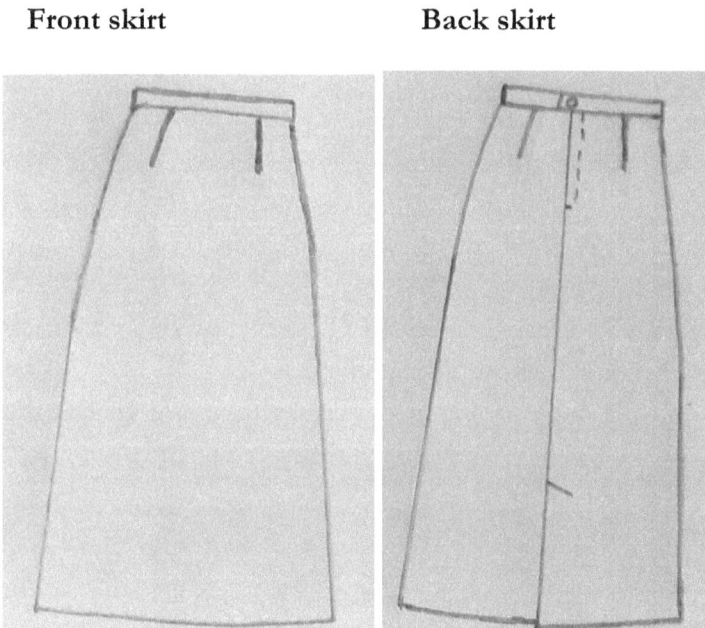

Measurements required to make a straight skirt are:

1. Waist measurement
2. Hip measurement
3. Shirt length

Preparation for drafting the skirt

(a) Waist measurements are divided by four and then 1" is added for seam and another 1" for the dart allowance for a skirt with darts. For example,
Waist is 30"

Therefore 30 ÷ 4 = 7.25 + 1" seam allowance and 1" foot dart = 10.25"

(b) Divide the hip measurement by four and add 1" for seam allowance.

Hip measurement is 36"

Therefore 36/4 = 9 + 1" seam allowance = 10"

(c) The shirt length measured includes the waistband. The finished waistband width is subtracted from the skirt length and then Hem allowance is added.

Skirt length is 27.

Therefore skirt length 27 minus 1 ½" waistband depth then add 1 ½".

Then add 1 ½" for hem allowance. Waistband width varies from 1" to 1 ½".

A more flared skirt will be added 1" hem allowance so that it does not become bulky around the hem.

The skirt has a zip opening at the centre back and vent at the centre back, back and front darts and side seams only.

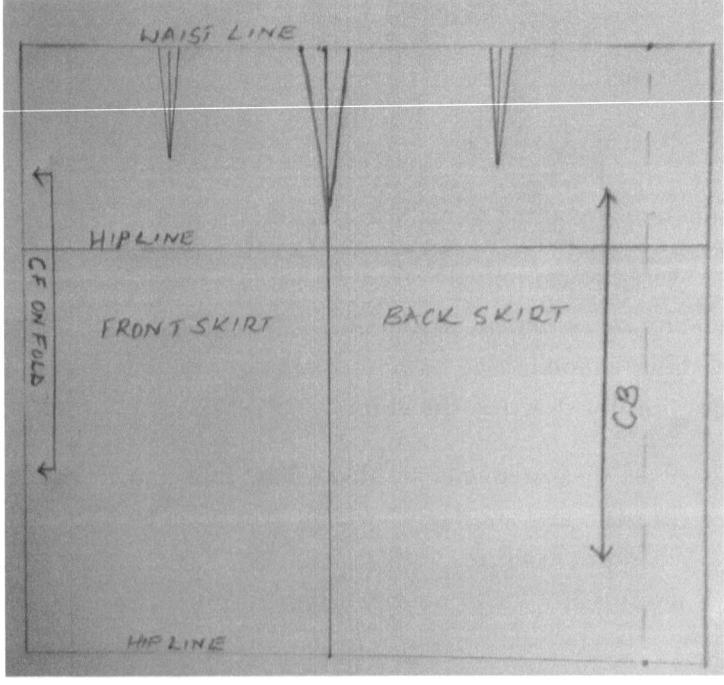

12.1.2 Front skirt
1. Front skirt will be cut on fold, fold fabric into two along a straight grain. Fold in such a way that the layer at the top is 10.25" which is the (largest) part (waist +2") so as not to waist the fabric, but fabric that is above 45" width can be folded twice since it can produce both the front and back skirt patterns.
2. With the fold line being the centre front, draw a line across the fold line and measure 10 ½" this will be the waist line of the skirt.
3. Measure Hip line 8" down from the new line along the fold line and square across on the second, hip line is normally 8"-9" from the waistline. Mark 10" from the centre front: which is the hip measurement + 1" for seam allowance.
4. Measure the skirt length 27" down from the waist line along the centre front and square across.
   Mark 10" + 1" for ease along the Hem line and connect the mark to the Hip line and then a slight curve from the hip line to the waistline along the side seam (open edges of the fabric).
5. Mark the seam allowance allowed (1") along the side seam. This will be the fitting line (stitching line).
6. Mark midway from the C.F. to the fitting line along the waistline to find the position of the darts. Mark at ½" on both sides of the midpoint which is the side of the dart and square down from the dart midpoint, measure 4.5" and mark the tip of the dart.
7. Cut out along the waistline, side seam and hem line of the skirt, notch the position of the dart and hip line. A notch is also made at C.F. both at the waistline and hem line.

12.1.3 Back skirt
The back skirt will consist of two patterns since it has a centre back seam that divides the back into two. Therefore, it should be cut on two layers of fabric.

1. Lay two layers of fabric from the remaining piece, and place the front pattern on the fabrics leaving 2" zip allowance from the edge of the fabric to the C.F. which will be C.B. of the back skirt.
2. Trace round the skirt, and mark the position of the hip line, dart and the C.B.
3. Remove the front skirt and cut and notch the position of the hip line, C.B. and dart on the back skirt patterns.

Waistband

On the straight grain fold the fabric into two of length waist measurement plus 2.5" for overlap and seaming allowance which will be 30 + 2.5" = 32.5".

Measure waistband width plus 0.5" for seam allowance across the fold line which will be 1.5" + 0.5" = 2" and cut out the waistband.

Skirt assembly

At this time you are required to neaten the side seams, back seam and hemline of the skirt patterns. This is done to prevent fraying and make your work look neat.

1. The starting with the front pattern, fold the R.S. together. Find the centre of one of the dart and stitch 0.5" from the waistline tapering to notching down 4.5" from the waistline. Repeat the same to all the darts on all the three patterns.
Press the front darts towards the side seams and back darts towards the centre front.
2. Place the back pieces R.S. together matching the C.B. and Hip line of the two patterns measure 7" down the C.B. from the waistline and 8" up from the Hem line for the zip and vent respectively.
3. Machine stitch between the two points leaving the position for the zip and vent unstitched and press the seams open.

4. Aline the zipper face down on the seam line. The zipper teeth will be in line with the centre of the seam.
   Pin the zipper in place.
5. Sew from the top of the zip down to the bottom starting from the left side and backstitch 3 times at the bottom and turn to the right side finishing at the top.
   Remove tacking pins and press the zip.
6. Place the front skirt on top of the back skirt, match the fitting line and stitch through the side seams and press both seams open
7. Attach interfacing to the Wrong side of the waist band and pin the Right side of waist band to the Right side of the skirt along the waist line, leaving an overlap for the button hole on the Right hand side of the back skirt and machine stitch the waist band in place.
8. Stitch the ends of the waist bands with the Right sides together and trim corners, turn the waist band Right sides out.
9. Fold the waist band over seamline to the Wrong side of the skirt and stitch along the bottom of the waist band along the waistline seam on the Right side of the skirt and press the waist band.
10. Mark the skirt hem line and fold the hem allowance to the Wrong side of the skirt. Stitch the hem in place all-round the skirt.
11. Cut the button hole on the waist band's overlap extension and neaten it with the button hole stitch.
12. Mark the position of the button on the opposite side of the overlap and sew on the button.
13. Press the skirt properly and cut all the hanging threads.
14. Cut the buttonhole on the waistband overlap and neaten it with the buttonhole stitch.

12.1.4 8 piece panel skirt

This is a skirt that consists of 8 equal pieces (panels). 4 for the front skirt and 4 for the back skirt.

All the patterns will also have the same shape. Similarly, the method used to create this skirt can be used for skirts for 8, 12 or more panels.

Front skirt design

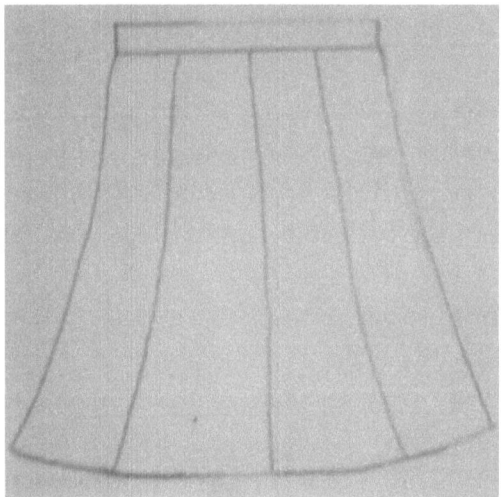

Back skirt design

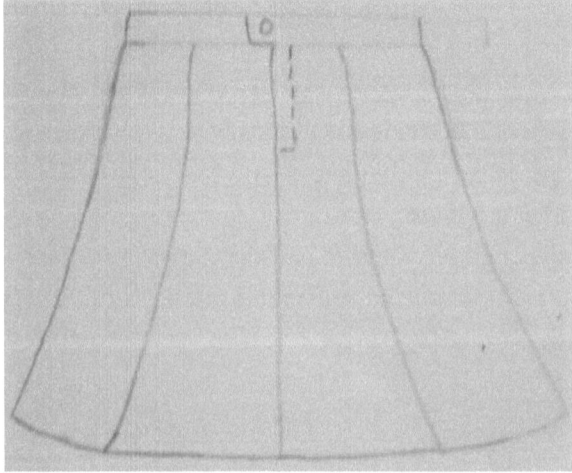

How to cut skirt

The skirt measurements are:

- Waist 32"
- Hip 42"
- Skirt length 24"

Square down and across on the fabric to make a "T" shape. The line drawn across will become the waist line. Then measure the hip line down 8" from the waist line. And square across on both sides. Measure the skirt length from the waistline and draw another parallel line to the hipline and waist line this will be the hemline. Now that the skirt grid has been drawn, the next step is to use the available body measurements to complete the outline.

The waist and hip measurements will be divided by the number of panels desired.

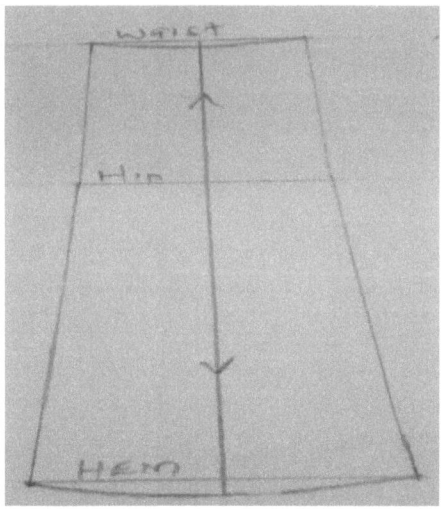

For instance, to create an 8 panel skirt of waist 32" [81.5 cm].

32" ÷ 8 = 4" [10 cm]

Add ½ " for seam allowance to make it 4 ½

Measure 4 ½ at the waist line with the vertical line being at the centre

The same will be done to the hip whose measure is 42"

42" ÷ 4 = 5 ¼

Add ½ for seam allowance =5 ¾ " [14 cm]

Hip measurements will be used on the hem line with the desired amount of flayer added.

Measure down ¼" on the centre of the waist line and curve the waistline. Extend the centre of the hemline down wards by ¼" and curve the hemline as well with the outside of the hem raised ¼ inch up.

Make the waistband to waist measurement plus 1 ½" and the width is 2 ½" (the finished width of waistband will be 1 ¼").

Skirt assembly

1. Neaten the raw edges of the eight patterns of the skirt. With the right sides facing stitch a pair of two panels at 0.5 inch seam allowance down from waist to hem line. Repeat the same to all the patterns. This will result to four pairs of patterns. Press all the seams open.
2. Take two pairs of the skirt panel and place one on top of the other with the right sides facing and matching the fitting lines. Machine stitch along the fitting line to complete the front skirt which by now has four panels. Ones again press the center seam open.
3. Repeat the same to the remaining two patterns but this will be stitched leaving the zip opening unstitched. Place the zipper facing down on the seam line. The zipper teeth will be along centre of the seam.
4. Pin the zipper in place.
5. Sew from the top of the zip down to the bottom starting from the left side and backstitch 3 times at the bottom and turn to the right side finishing at the top.

6. Remove tacking pins and press the zip.
7. Place the front skirt on top of the back skirt, match the fitting line and stitch through the side seams and press both seams open.
8. Attach interfacing to the Wrong side of the waist band and pin the Right side of waist band to the Right side of the skirt along the waist line, leaving an overlap for the button hole on the Right hand side of the back skirt and machine stitch the waist band in place.
9. Stitch the ends of the waist bands with the Right sides together and trim corners, turn the waist band Right sides out.
10. Fold the waist band over seamline to the Wrong side of the skirt and stitch along the bottom of the waist band along the waistline seam on the Right side of the skirt and press the waist band.
11. Mark the skirt hem line and fold the hem allowance to the Wrong side of the skirt. Stitch the hem in place all-round the skirt.
12. Cut the button hole on the waist band's overlap extension and neaten it with the button hole stitch.
13. Mark the position of the button on the opposite side of the overlap and sew on the button.
14. Press the skirt properly and cut all the hanging threads.
15. Cut the buttonhole on the waistband overlap and neaten it with the buttonhole stitch
16. Mark the position of the button on the other side of the overlap and sew on the button.

12.2 Blouse

The following measurements are required to draft a good bodice for the blouse. The bodice would be developed into any blouse design

a) Shoulder
b) Bust
c) Length

d) Sleeve length
e) Nape to waist
f) Hip

Front blouse

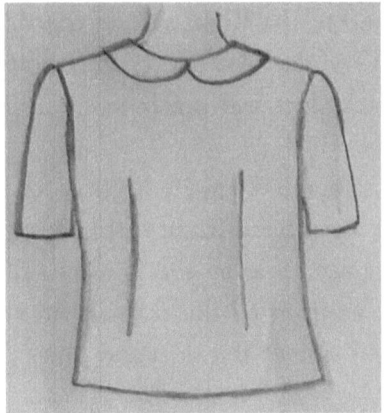

Back bodice

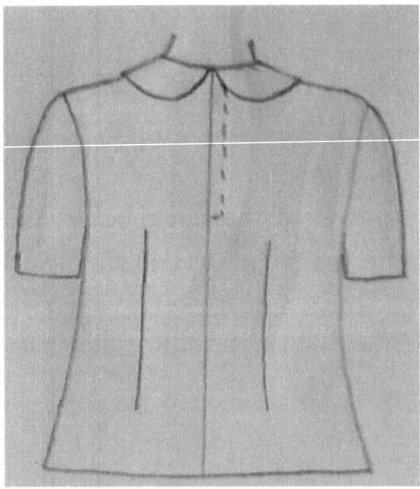

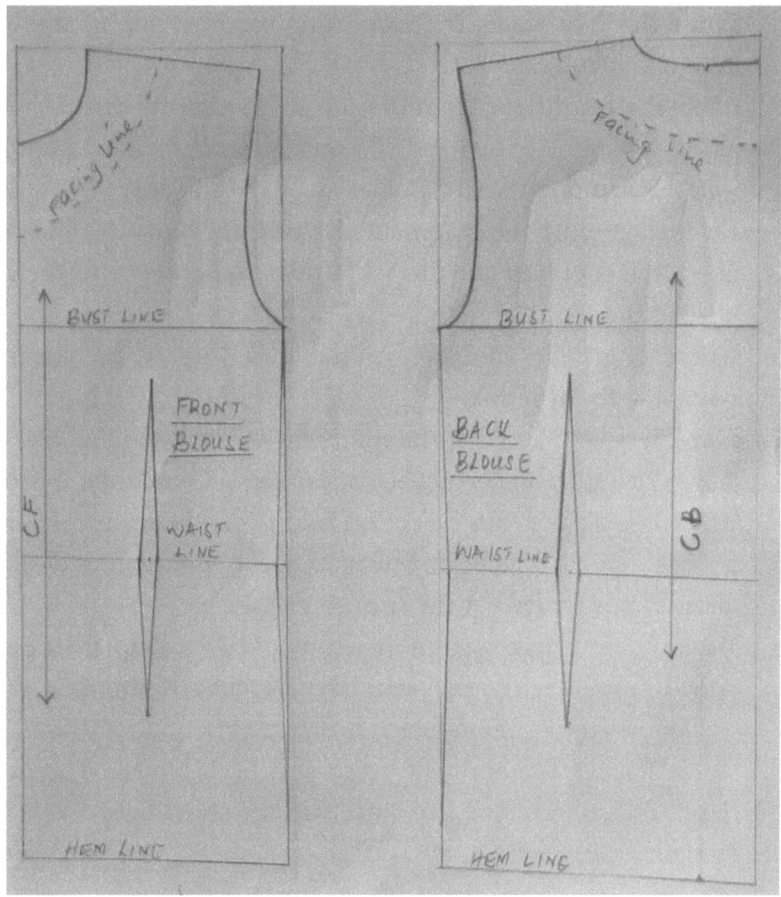

How to cut

1. Square across and down from the top left corner. Measure ¼ busts from the top corner down wards to locate the bust line, square across and mark ¼ bust plus 1" for seam allowance along the bust line.
2. The waist line is nape to waist measurements from the shoulder line and square across, mark ¼ waists plus 1" seam allowance and 1" dart along the waist line.
3. Square cross the hipline at 8" down the waist line square across and mark ¼ hip plus 1" seam allowance along the hip line.

Draw the side seams by connecting the bust line to the waist and hip line.

4. Mark the front neck from the top left corner by marking $2^1/_2$" along the shoulder line and 3" down along the centre front line; join the two point's neck point with a smooth curve.
5. Divide shoulder measurements by two and add 0.5" and mark along the shoulder line drop 1" down and connect to the neck edge at the shoulder.
6. Curve from the shoulder to the bust line at side seam to complete the arm hole.
7. Darts will be marked at the midpoint of the waist line, the dart size is one inch and 5" up the waist and 4" down the waistline. The front bodice may be cut on fold for a blouse that will have a back opening while it will cut and a button stand or zip allowance included for the front opening.
8. Back blouse is traced from the front and 2" for zip allowance is added from the centre back. The neckline is dropped by 1" unlike the front pattern which is dropped by 3".
9. Trace facings for both the front and back blouse. The facing depth is 2.5" from the neckline. Facings should also be cut on straight grain.

Collar

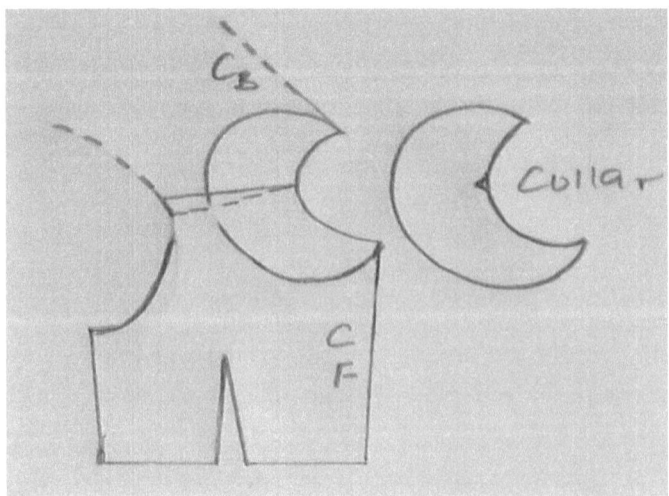

Place shoulders of back bodice to shoulders of front bodice neck points touching and overlapping by ½ inch draw collar shape. Trace out and cut four pieces.

Sleeves

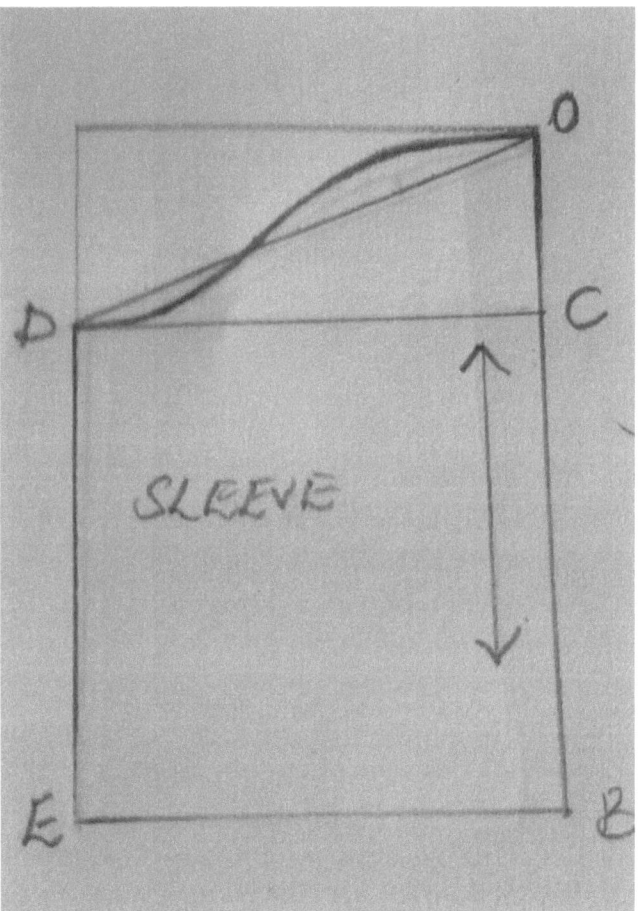

You will need the sleeve length and bust measurements to be able to draft a sleeve. The sleeve will be drafted on fold cut two pieces.

Square down and across from point O, mark point C is $1/_{12}$ Bust and squire across to D ¼ bust, B is sleeve length + 2".

Join O to D with a straight line, mark point E across from point B which is ½ sleeve width plus 1" for seam allowance and join point E to D.

Curve the sleeve armhole raising 1" at mid-point.

Blouse Assembly

Ensure you have all the patterns namely:

    a) 1 pc Front Blouse
    b) 2 pc Back Blouse
    c) 2 pc Back neck facing
    d) 1 pc Front facing
    e) 2 pc Sleeves

1. Start by stitching the double pointed darts by folding the darts Right sides together matching fitting line the. Machine stitch along the fold line three stitches. Follow the fitting line to 0.5" at the waistline and tapering to nothing towards the other end and run 3 more stitches along the fold line to secure the dart.
2. Repeat the same to all the darts and press from the front darts towards the side seams and back darts towards the centre back.
3. Place the Back patterns on top of one another Right sides facing and matching the C.B. fitting line. Measure the zip opening according to the length of the zip being inserted from the neckline downwards. Machine stitch the remaining part of the Blouse to leave the zip opening. Press the seam open.
4. Insert zip at the C.B. Place the folded edge against the zip teeth and stitch along the folded edge from the bottom of the zip to the top.
   Place the folded edge on the stitch line of the opposite side and pin starting at the top of the zip stitch downwards and across on seam line.
5. Matching the fitting lines at the shoulders place the back and front patterns with the Right sides facing, machine stitch the shoulders and press the shoulder seams open. Stitch together

the front and back facing at the shoulders and finish the outside row edges.
6. Place the two pieces of collars Right sides together matching fitting lines and stitch outer edge of collar.
7. Trim turning to 0.5 cm and snip notches into all cornered turnings, turn collar Right Side out bring stitch line on to the fold and press carefully.
8. Place underside of collar to the Right Side of garment and bring edges to meet at Centre Front and Centre back lines. Place the facing on top of the collar, match notches and the shoulder seams and pin in place. Machine stitch round the four layers of the fabric.
9. Clip the raw edges on the neckline to U-shape, turn to the Wrong Side and under stitch. Turn collar to the Right Side and press.
10. Turn the blouse right sides together and matching the fitting lines of the side seam, stitch the side seams from below the armhole to the hem line and press the seams open.
Machine stitch the sleeve seam and press it open, turn the sleeve to the right side.
11. Insert the sleeve into the armhole of the blouse with right side of sleeve to the right side of the blouse, match the underarm seam and the sleeve head matched to the shoulder and pin.
12. Machine the sleeve in starting at the under arm seam, stitch so that the sleeve is at topmost and stitch straight over the shoulder and overlapping at the under arm to reinforce the stitching.
13. Turn up the hem for the sleeve matching the seams together and stitch the hem in place. (The hem can be machine stitched or worked by hand) and repeat the same procedure for the blouse.
14. Cut all the hanging threads and carefully press the blouse

12.3 Gents Shirt

The shirt is an important garment. It is in use for almost all the time of the day and so is required to be replaced by a new one in a short period thus creating a great demand for shirts.

Front shirt   Back shirt

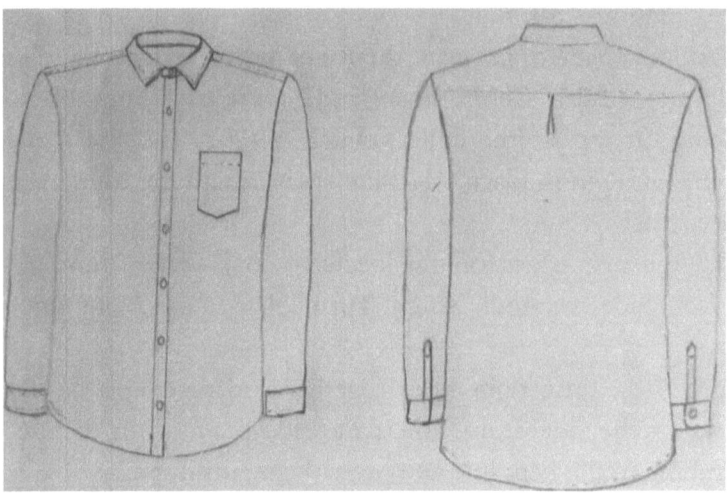

Measurements required to produce shirt patterns are

        i   Breast/chest
        ii   Shoulder width
        iii Sleeve length
        iv Waist length
        v   Shirt length

Before drafting fold the cloth in such a way that two folds of two layers each will form the front and back shirts.

Front shirt pattern & Back shirt pattern

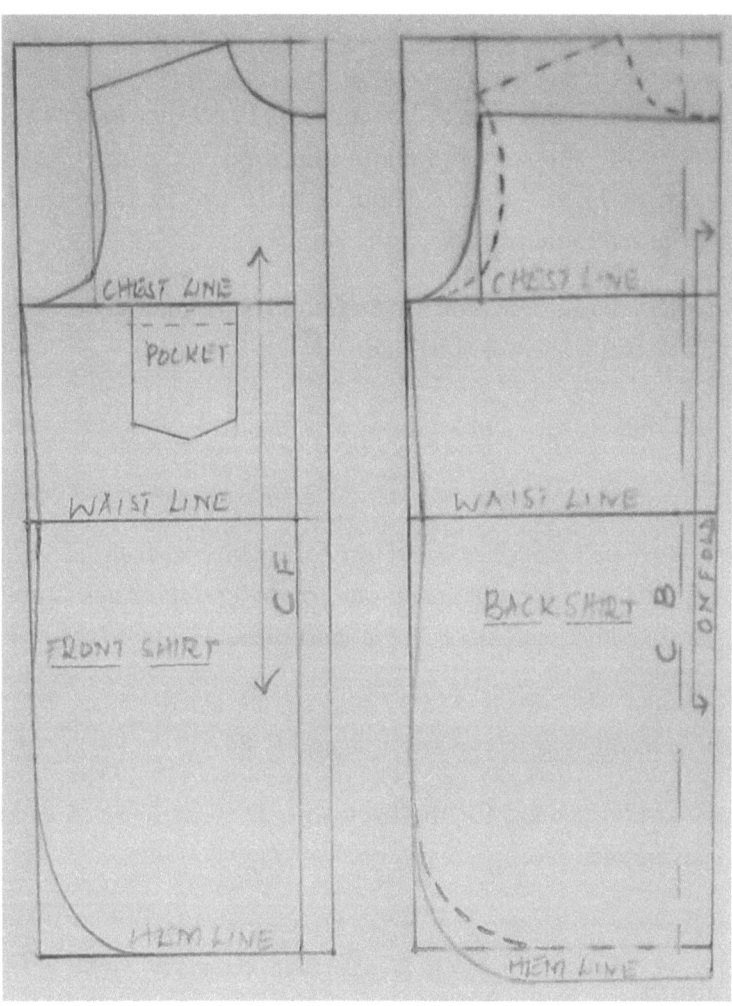

Front part

Square down and across, shirt length 1" and one-fourth breast plus 2" across respectively. Measure ¼ breast from 0 at the shoulder down wands to locate the chest line and square across. 0 to waist length and square across.

Measure $^1/_6$ neck across and down on the shirt centre front from 0 to 9 on the shoulder line and down along the centre front, and shape the neck line with a smooth centre.

Measure shoulder width plus ½" from the C.F. at 0 across to 13. Drop 1½" to create the shoulder slope and join the slope from 13 to 9 and square down to 14 at the breast line 14 to 16 up the from the breast and curve the arm hole from 15, 16 to 4.

Shape in the waist at the side by ½" and curve in towards the C.F. from midway from waist to shirt hem.

Cut out the shirt.

Back shirt

The back shirt pattern will be cut on fold with the back pleats place long the fold line. Use the front pattern to cut the back shirt. The button stand will be equivalent to the shirt pleats.

Measure down 3 ½" from 0 to 17 and square across, curve ¼" from the armhole to midway of the armhole and C.B.

Adjust the armhole curve for the back shirt at 16 by 1" so as to have a slightly curved armhole.

Yoke

Yoke pattern

Square out lines with the fabric. On fold measure 3 ½" along fold line, and mark 22 and square across from 21 to 23 shoulder width plus ¼ square down to the line from 22 to meet at 24.

Measure across ¼ neck from 21 to 25 and 1 ½" up from 25 to 26. Complete the Back neck curve from 21 to 26 with a slight curve.

Rise the shoulder at 23 by ¼" and outwards by ¼" and 24 up by ¼" then complete the yoke by joining 24 to 23 and 27.

The yoke should be cut in two lays.

Collar

Collar pattern

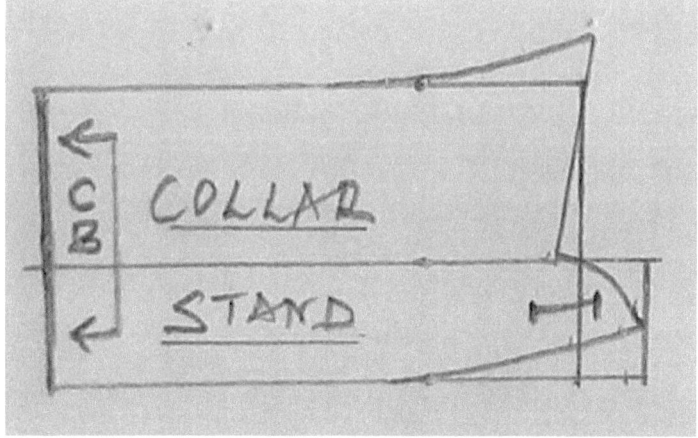

The shirt collar will be cut of fold and laid of two layers of fabric.

Measure along the fold line 1 ½" for the stand 0-1 and 1 ¾" for the collar 1-2.

Square across from 1-3 half neck size 3-4 is 1 ¼" for button stand. Square from 4 to 5 along the line from 0. Shape the stand from 4-5 with a curve.

Square down from 2 ½" from 3 to 7 and square across 1" toward the C.B. from 7-3 and curve to two on the fold line to create the fall.

Cut out the collar and separate the collar and the stand.

Sleeve

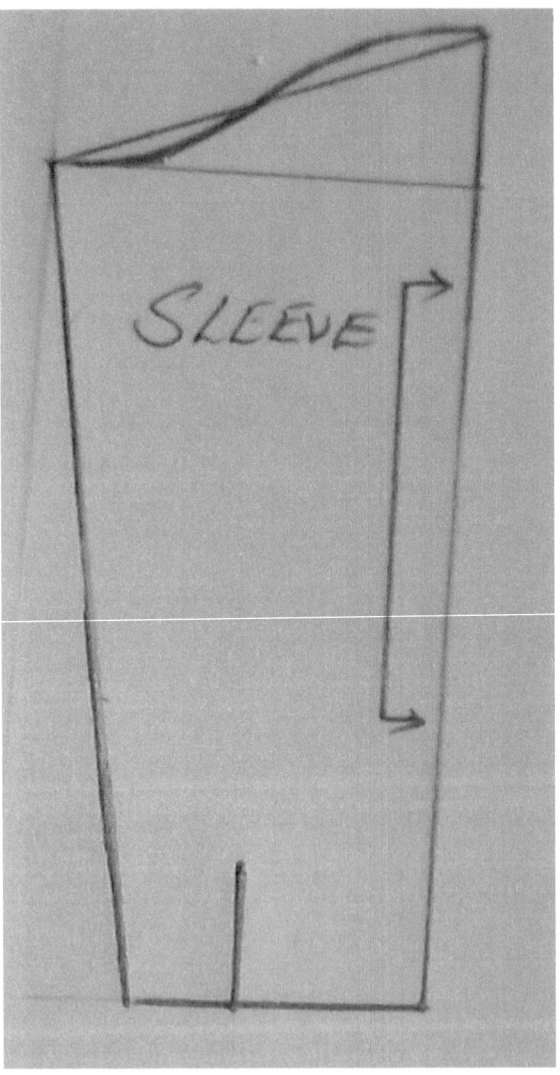

Fold the fabric into two while drafting the sleeve. The sleeve head will be on the fold line.

From 0 Square across ¼" breast less ¼" and down to 1 sleeve length minus 2 ½" cuff width + 1 ½" for seams.

Measure one-twelfth down from 0 and square across to 2 and join 0-2 with a straight line square across from 1 to 3 one sixth breast measurement and join 2-3 with a straight line.

Curve the sleeve from 0-2 raising up at midway by 1".

Cuff

Draw a rectangle 3" x 5". To be cut in four pieces two for each sleeve.

Shirt assembly
Preparing the shirt front opening
On the right pattern of the shirt which the buttons will be sewn on press the extended facing to the wrong side with the raw edges turned under, edge stitch the fold and press.
Take the button stand strap and place it on the wrong side of the left opening with the right side of the strap facing the wrong side of shirt. Stitch the strap in place and turn the strap to the right side and press. Edge stitch 0.5cm parallel to the seamed edge.
Turn the raw edges of the other end into the wrong side press and machine stitch 0.5 cm along the fold.

Sew the shirt yoke facing

Tuck the shirt pleats in place. Take the yokes, Sandwich the back shirt between the two yokes pieces, wrong sides facing out and notches matching. Pin in place, and then stitch long the horizontal seam a 3/8" seam allowance, Trim the yoke seam and turn the yokes right side out press, and top stitch on the yoke 0.5 cm

With the wrong sides of the front and the back shirts facing match the shoulder seams for the yoke facing and the front shirt and machine along the shoulder seam repeat the same with the other front shirt. Roll the shirt back upwards in a tight roll and leave it there. Roll the shirt fronts toward the shoulder seam so that seam is now exposed and you can match the shoulder seam to the yoke facing shoulder seams as seen above. With front shirt sand witch between the two yokes sew yoke and facing at shoulder seam. Pull the shirt in place and press the shoulder seam, top stitch the shoulder seam on the yoke 0.5cm form seam

How to Sew a Shirt Collar
Add Fusible Interfacings to Collar Pieces

The shirt collar will be made up of two pattern pieces: the collar and the collar stand. Attach fusible interfacings to one of each piece to give them structure and body. Transfer the pattern markings — on the collar stand

Sew the Long Collar Seam

Once your interfacing is done, with the markings transferred, move the pattern pieces out of the way and place the collar pieces right sides together. Sew the seam on the long outer edge, it's recommended that they are trimmed down after stitching to reduce bulk.

Trim and Turn

Carefully trim away the point of the collar. Depending on the angle of your collar point, you may have to trim more than a diagonal. Regardless, the goal is to have the seam allowances Turn the collar right side out. With a point turner (or another blunt tool), gently push the corner to create the point. Take your time so you don't poke a hole in the fabric; you can tease the last bit of the corner out with a pin or needle if necessary.

Press the collar flat. Roll the seam edge ever so slightly toward the underneath, as it offsets the bottom edge just a bit.

Attach the Collar to the Collar Stand

Take the collar stands, Sandwich the collar between the two collar stand pieces, wrong sides facing out and notches matching. Pin in place, then stitch from corner to corner using a 3/8" seam allowance.

Trim the Seam Allowances on the Collar Stand

Trim the edges of the now-attached collar stand, leaving about 1/8", There's no need for notches, and the seam turns right side out beautifully.

Press Your Collar

Turn your collar and stand to the right side and press the round sides, working the seam to the edge to get a crisp finish. Press the rest of the collar.

Patch pocket

From the remaining fabric cut a pocket 6.5" long by 5" wide, turn the top edge over to the wrong side. Turn 0.5" of the raw edges of the top edge and tuck 0.5cm of the raw edges of the pocket under to the wrong stitch the top edge in place and press. On the left front shirt Place the pocket 2" from the center front with the pocket mouth along the breast line and tuck the pocket in place. Start by reinforcing the pocket. Machine stitch 1mm from the edge starting from one corner of the pocket finishing with reinforcing the other edge of the pocket.

Placket opening

Place the piece of fabric for placket on the shirt sleeve with the right side of the placket facing the wrong side of the sleeve and pin it in place. Machine a rectangular box at the position of the opening 0.5" wide and equal to the length of the opening. Slash the placket straight down the center between the raw of stitching. Turn the placket to the right side of the sleeve and press. A rectangular gap with sharp corners will be formed. Fold back the long edge of the shorter side of the placket. Place the folded edge on top of the stitch line, pin in place and machine the folded edge.

Fold the other part of the placket across the shorter side and press, fold below the top pointed end following the cut edge and press.

Machine stitch the longer folded edge of the placket in place while you ensure the underside of the placket is not attached as well. Continue machining around the pointed edge and stitch an X on the pointed area.

Attaching the sleeve

Will use the flat method to attach the sleeve, place the right side of the sleeve on right side of the shirt ensuring that the placket is towards the back of the shirt. Machine the sleeve to the shirt starting from the front shirt finishing with the back shirt. Repeat the same with the other sleeve. Press the armhole seam towards the shirt.

Trim the shirts seam allowance by half and fold the sleeves seam allowance turning over to the reduced seam allowance press and machine 1mm from the edge do these to the other sleeve as well.

With the shirt turned to the right side match the under arm seams and the shirt side seams machine the seam 0.75" press the seam allowances open and trim the back seam to 0.25" and press front seam to the back side and fold the seam allowances to the wrong side and machine to complete a double stitched seam.

Cuff

Apply interfacing to the top cuff. Place the top cuff on top of the under cuff with the right sides facing and machine along the lower seam. Press the seam open, turn the cuff right sides out and press

Take the part of cuff that's not interfaced, place it on the wrong side of the sleeve facing the right side of the cuff and machine stitch. Pleat the sleeve for it to fit into the cuff. Turn the cuff right sides together and stitch shorter ends in line with the placket. Turn the cuff to right side and push the corners out.

Turn under the raw edge of the interfaced cuff in line with the seam of the sleeve machine the end in place, top stitch 0.5cm round the cuff.

Hemming the shirt

Turn raw edges under to the wrong side of the shirt and edge stitch the fold line.

Button and button holes

After hemming, it's time to make button holes. Gents shirt button holes are made on the left side while they are made on the right side for ladies. The first button hole is made on the collar stand and the rest on the shirt button stand at an interval of 4". The button holes should be equal to the diameter of the buttons and should be neatened. Buttons should be stitched exactly opposite the button holes on the right side of the corresponding part.

12.4 Trouser
Pleated trousers

These trousers have waistband and pleats. Due to the pleats, more fullness is obtained in the seat region. The waistband is extended by about four inches and is used in adjusting the waist-ease by hook or button. Do not make economy in taking pleats. At least the first one should be large enough.

Pleated trouser design

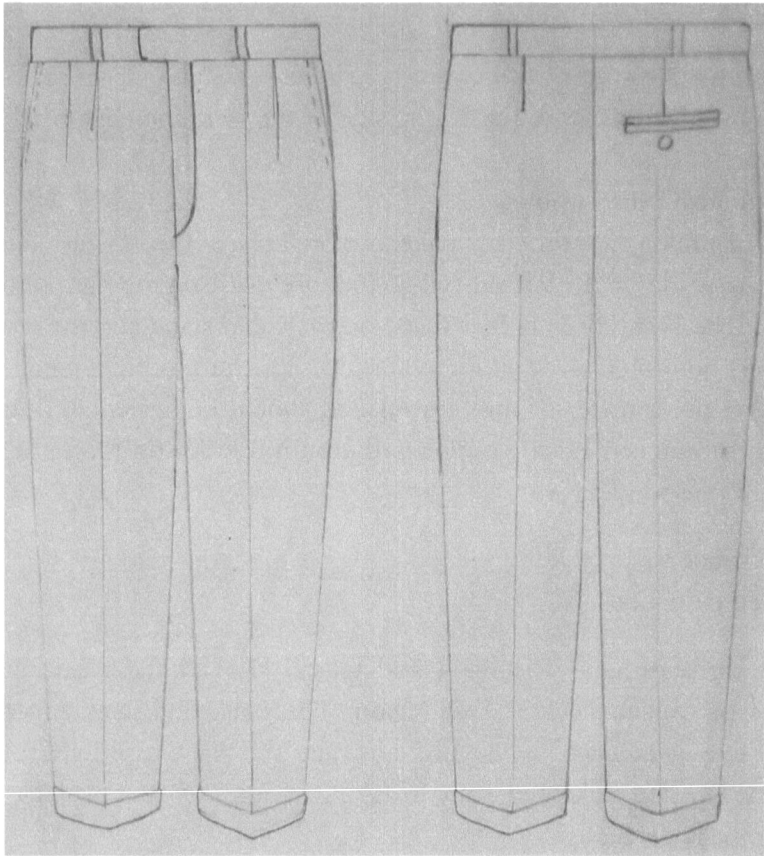

Measurements for drafting a pair of trousers:

a) Full length – 40"
b) Leg length – 28"
c) Waist – 30"
d) Seat – 36"
e) Knee – 23"
f) Bottom – 21"

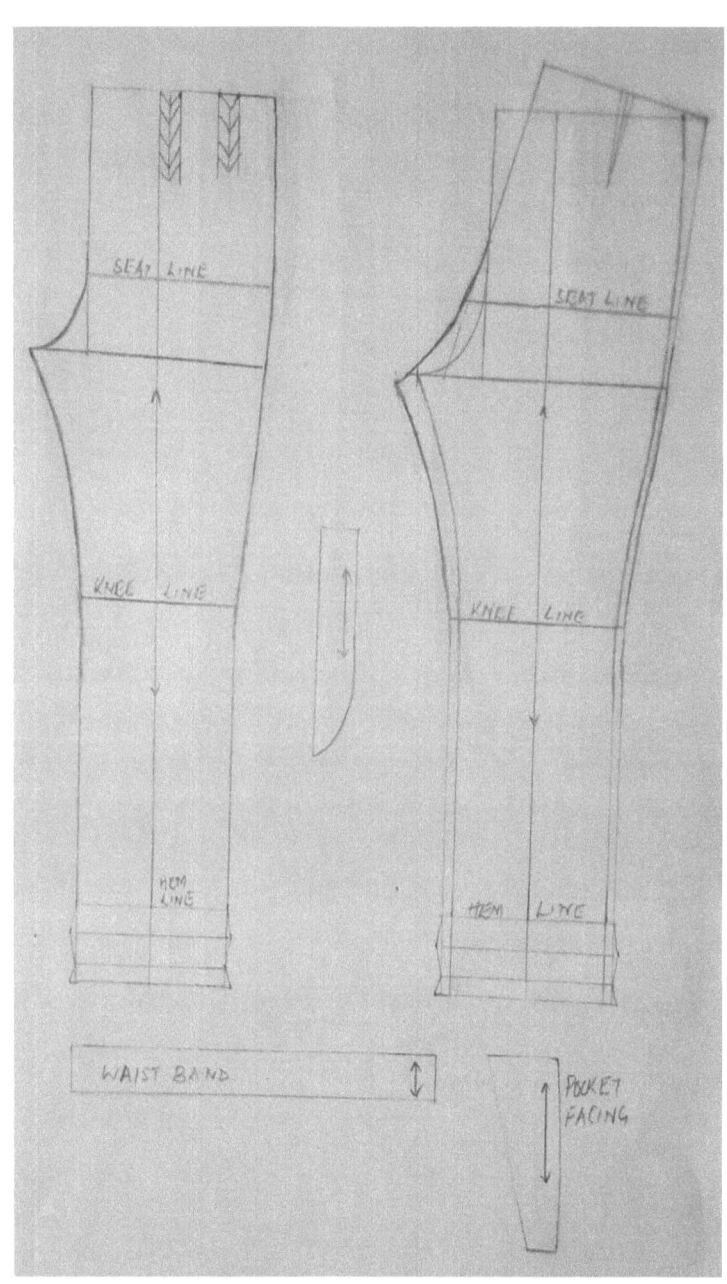

Front part

1-0 = full length less 1 ¾" for the waist-band and + ¼" for a seam (38 ½").

2-1 = leg length (28)

Square out lines from 0, 2 and 1.

3-2 = one-third seat + ¼" (12 ¼)

4-2 = one-fourth seat (9).

Square up from 4 to 5.

6 is midway 2 and 3.

Square up and down from 6 to 7 & 8.

9-5 = one-fourth waist + 3" or as required for the pleats (10 ½").

10-7 = 2" for the first pleat.

11 is midway 9 and 10.

Take the second pleat at 11 one inch wide.

12-4 = one-sixth seat (6).

13-4 = half of 3-4 + ¼" (1 ¾").

Shape the fly 12-13-3 as shown.

14-4 = one-twelfth seat (3).

Square out from 14 to 15.

16-6 = half of 8-6 less 2" (12).

Square out from 16 to 17 & 18.

17-16 = one-fourth knee less ¼" (5 ½).

18-16 = one-fourth knee + ¼" (6). This ¼" is taken extra to balance to looseness of the pleats.

19-8 = one-fourth bottom (5 ¼).

20-8 = 19-8 (5 ¼). shape the leg-seam 3-17-19 as shown.

21-3 = ¼". Give the dress-shape as shown by dash line, hollowing ¾" at 13. Shape gradually in the leg-seam, 7" below 21.

Dress is a sartorial term. The man generally dresses on the left side means he wears the trousers in such a way that his genitals (sexual organs) remain on the left side remains comparatively empty. So the cut (dress) is taken in the right side in order that both sides should appear equal after wearing.

Due to the dress-cut, a twisted front-crease results so in modern loose trousers, dress-cut should be omitted. Even the draft shown here is loose enough to dispense with the dress-cut.

22-15 = 1 ¼". This looseness is taken in the seat region in order to balance the looseness of the pleats.

23-2 = as in the force of shaping from 22 to 18.

Shape the side-seam 9-22-18-20.

24-8 = 1 ½" or as the width of turn-ups.

25-24 = 24-8. 26-25 = ¼" for turning.

Back part

Cut the front part and place it suitably on cloth to draft the back part.

Join 21 to 12 and produce the line to 28.

28-7 = from 1" to 2". Take less in flat-seated figure and more in prominent-seated figure and more in prominent-seated one.

29-28 = ½".

30-29 = one-fourth waist + 2" (9 ½).

31-3 = 1 ½". 31 is 3/8" down the fork line.

Shape the seat-seam 29-12-31 hollowing ¼" from line 21-12.

32-12 = one-fourth seat + 2" (11).

33-17 = ½". 34-18 = ½".

35-19 = ½". 36-20 = ½".

Shape the leg-seam 30-32-34-36.

Leave ½" on both sides of the turn-ups also.

37-30 = 3". Take a dart ½" wide & 4" deep.

Waist band

1-0 = 2".

2-1 = half the waist + ½" (15 ½).

3-2 = 0-1. 4-3 = 4 ½".

5-3 = ½". 6-5 = 0-1 less ½".

7-0 = 1 ½" (on the right side only)

9-0 = 3" to 5" as per taste (on left side).

Allow inlay for turning as shown by wavy lines. The band is shown on canvas so allow seams on all sides. Cut the right side on line 7-8, and the left side on line 9-10. If the extension of band is not required cut the left side on line 0-1.

As 1" seams are allowed in the draft, sew ¼" inside the chalk-line everywhere.

Front part

Measure down from 0-1 full length less 1 ¾" for waistband 0-2 10". Square across at 0, 1 and 2.

Trouser assembly

Side pocket

Apply a fusible interfacing on the wrong side of the front trousers approximately 1 inch along the side pocket opening.

Place the pocket lining for the pocket bag to the Right side of the front trouser matching the stitching line. Stitch the lining in place with 1.5 cm seam allowance. Open up the pocket and press the seam towards the wrong side

Turn the lining to the wrong side and press so that the lining is not seen on the right side, top stitch the pocket opening ½ cm from the edge.

On the right side of the trouser pin the pocket facing on the lining along the pocket mouth. Match the seams and pin in place.

Attach the pocket facing on the pocket bag and stitch the pocket bag together using a 1.5 cm seam allowance.

Neaten the raw edges and the seam allowance around the pocket. Neaten the side seam allowance of the trouser and make sure the fabric lies flat where it joins on the side seam. Repeat the seam to the other hip pocket.

Stitch the back darts to a depth of 2.5" and 0.5" width and press the darts towards the centre back.

Inserting the zip

Stitch the seam, leaving the gap for the zip; mark the centre front lines on the two patterns.

Fold left and front along the centre front line. Place the fold next to the zip teeth and pin in place. Machine stitch along the fold line extending past the seam stitching line.

Bring together the overlapping flaps and pin them together. Turn the trouser to the wrong side and carefully open the flaps. (You should have the flap is left unattached to the left side with zip pinned on top). Ensure you do not remove the pins on the right side.

Stitch the zip on to the left flap and turn the trouser to wrong side and attach zip to single layer of right flap.

Turn trouser to the right side once more and stitch the curve line to join the flaps and give the fly its style. On outside overstitch the curve line.

Back pocket

This is stitched on the Right Side of the back trouser. The kind of pocket made is a jetted pocket.

The main components are welts and lining for the pocket bag and the pocket facing.

First make the welts of length 6" by applying fusible interfacing to the wrong sides of the two strips of fabrics, fold in half lengthwise towards the wrong side and press.

Place the R.S of the welts to the right side of the back trousers with the fold lines facing away from the pocket mouth position on both the upper and lower positions.

Machine stitch the two welts in place. Make sure the two rows of stitching are 1.25 cm apart and parallel to each other and exactly the same length.

Take the lining and press in half, right sides facing over the welts, the crease line on the lining should be placed between the two welts.

Working from the wrong side stitch the lining in position over the stitching lines that are holding the welts in place slash through just the lining along the pressed crease line.

On the opposite side slash through the trouser and not the welts. Slash into the corners right to the stitching lines.

Pull the lining through the slash to the wrong side of the trousers, push through the ends of the welts.

Stitch across the welts and the triangle and around the pocket.

Press everything in place.

Having made the fly and all the pockets it is time to join the under trousers to the top trousers, place the right sides facing together matching the side seams and the fitting lines for the side seams and machine stitch from the waist to the hem of the trouser including hip pocket linings attached to the front trouser and press the seams open. Repeat the same procedure to the side of the trouser.

Cut the fabric strips 1 ¼" wide and long enough to make six belt carriers of length approximately 2".

Press the long edges of the fabric carriers to the centre, wrong side to wrong sides and press the carriers.

Machine along the centre of the carrier securing the edges. Cut the belt carriers into desirable sizes.

Starting at each side seam place the carriers to the waist at regular intervals on the right side of the trouser and stitch to secure the carriers.

Apply interfacing to the wrong side of the waistbands, with the right side of the waistbands, with the right side of the waistband facing place the waistband on the trousers around the waist and stitch the waistband in place with the carriers sandwiched between the trouser and the waistband.

Press the waistband and the carriers away from the trouser. Repeat the same on the other part of the trouser, place the waistband onto the right side of the fabric waistband with the other side of the waistbands in between them and stitch the waistbands in place turn the waistband towards the seam allowance and edge stitch on the wrong side waistband attaching seam allowance for the waistbands with it. Turn the unstiffened band to the wrong side and place the hook overlap over the underlay and mark the position for the hooks. Put the hooks in place and turn the waistband right side facing and stitch to complete the squared end of the band and the straight edge as well.

Turn the trouser to the wrong side and match the back seam ensuring that the trouser has the right waist measurements. Match the waistband seams and machine stitch the trouser back seam from the waistband down to the folk seam and press the seam opening.

Turn the unstiffened band to the wrong right and ensure that all raw edges are enclosed. Complete the waistband by machining between the waistband and the trousers starting from one part of the opening all through to the other and press the waistband in place.

Turn the trouser to the wrong side and match the seams from the back and front, inside leg seam allowance and fitting lines and knee position. Complete the inside leg seam from one part of the hem one on the other side of the leg.

Press the seams open and turn the trouser to the right side out.

Measure the trousers full length and mark the hemline on both legs. Turn the hem allowances to the wrong side and press. Stitch the hem in place using the hemming stitch.

Cut all the hanging threads and remove all the tacking stitches. Press the crease lines and trouser appropriately.

12.5 Revision Questions
1. List the measurements that are divided by four while drafting a straight skirt
2. State the measurements added to the skirt front waist while drafting a skirt with waist darts
3. What determines how the waist and hip measurements are divided while drafting a panel skirt?
4. What is the adult's measurement for hip to waist?
5. Outline the order for assembling a bodice with a faced neckline
6. State areas where interfacing is applied on a blouse
7. Highlight the patterns required for long sleeved shirt
8. List shirt patterns where interfacing is applied
9. Name the type of seams that are used on a shirt side seam
10. List the patterns required to stitch a trouser
11. List the trimmings applies on trousers
12. Outline the procedure for assembling a trouser

Trade Terms

American Lapels: Pointed lapels, D.B. Lapels

English Lapels: Step lapels; S.B. Lapels; V Shape lapels

Basting: the name of a kind of sewing in which long stitches are used just to hold the two parts in place for a short time.

Bearers: these are used in trousers and breeches. They are the pieces of cloth fastened to one of the side-seams which are then buttoned towards the other side-seam in other to bear the weight of trousers when the fronts are unbuttoned, and also to fill up the parts cut away in making the falls.

French bearers: these are used in trousers. A strip of cloth is attached to the fly-catch (on the right side) which is fastened towards the left side-seam with hook or button from the inside (not visible from outside).

Bridle: a narrow strip of cloth which is padded on the canvas along the crease-line of the lapels in coat-making.

Dungaree: the name of a rough cotton cloth of blue colour. A garment made of this cloth for firemen, workmen and a play or work dress of blue colour for boys is also called Dungaree.

Falls: instead of ordinary fly, these falls are made in trousers or breeches which serve the same purpose of fly (i.e. to make water). There are two kinds of falls: (1) Whole Falls, and (2) Split Falls.

Whole Falls: these are made in breeches (rarely in trousers too). They are made instead of ordinary fly but serving the same purpose. The front-piece falls down when unbuttoned (to make water) and is then buttoned up to the bearers.

Split Falls: these serve the same purpose as of Whole Falls, but the difference is that of a cut or split in both the fronts running vertically.

Fish: a cut or dart is taken in various garments to fit the waist section. As the shape of these cuts roughly resembles a fish, the same has taken root.

Fly: an inner flap to conceal the button-arrangement in trousers, overcoats, etc. in trousers it is attached on the left side and button-holes are made in it.

Fly catch: it is also called button-catch. It is attached to the right side of trousers and buttons are sewn on it.

Fork: in trousers the fork is point where the two legs join. It is also called Crutch or Crotch.

Jetting: a narrow strip of cloth used in finishing the mouth of cut-pockets. It looks like piping at the mouth of cut-pockets.

Jigger button: a name of a button which is attached inside the left overlap of a double-breasted coat to keep the underneath overlap in position. In ladies' coats it will be on the right side.

Mercerized: the cotton materials after going through a chemical process to become more lustrous, strong and fully shrunk.

Selvedge: a very narrow strip woven on both sides of cloth to prevent fraying and to strengthen the edges.

Sizing: a finishing process made on yarns and cloth to give strength, stiffness, and smoothness.

Welt: a strip attached to an edge to strengthen or adorn it. Welt is used in outside breast-pocket of coats and in waistcoat-pockets.

Petersham: Stiff ribbon used to reinforce waistbands, set inside top of skirt.

Additional Revision Questions

1. Why does one need a box or a container in a dressmaking workshop?
    A. To keep garment brought for repair
    B. Storing garments while under construction
2. Which statement is true about silk?
    A. The hot iron box leaves marks on the surface of the fabric.
    B. Silk fibers are inelastic
3. Choose the correct definition of pressing
    A. Pressing – the iron box is gently but firmly pressed down, lifted and pressed down again on to the next section of the fabric.
    B. Pressing – the iron box is moved back and forth on the fabric with little pressure being applied.
4. Which one of the following is not a method of marking out patterns
    A. Use of pinking shears
    B. Use of tailor's tacks
    C. Use of tailor's chalk
5. Which one of the following factors is not necessary to consider when buying a sewing machine?
    A. Price
    B. Can be serviced easily
    C. Country of origin
    D. Availability of spare parts
6. Select what not to consider when choosing clothes for good grooming
    A. Career
    B. Colour
    C. Style
    D. Type of fabric
7. Choose features to avoid for a tall and angular figure
    A. Well-tailored trousers

B. Double breasted coats and jackets
   C. Outfits in one colour, clinging fabrics
8. Choose the details to avoid for a short and slim figure
   A. Choose simple styles taking the attention from the waistline
   B. Small prints
   C. Skirts printed into waist
   D. Large prints
9. The following are the points to consider when making seams except
   A. Match balance marks and fitting line
   B. Tack firmly along fitting lines and remove pins
   C. Colour of the fabric
   D. Stitch exactly against the fitting or tacked lines
   E. Press each stitched line first and then press turnings open or to one side as necessary
10. Choose outfits which a French seam is used
    A. Shirt, jacket, panel skirts
    B. Children's wear, fine blouses
11. Which is the right amount for a pleated skirt?
    A. Three times the finished hip measurement plus turning
    B. Three times the finished waist measurement plus turnings
    C. Four times the finished waist measurements plus turnings
    D. Two times the finished waist measurements
12. Which statement is true on the reason tucks are used on children's wear
    A. Hold extra fabric allowed for growth
    B. To make children feel comfortable
13. Choose an area where gathering is not applied
    A. Side seam
    B. Waist
    C. Yoke

D. Sleeve head
E. Wrist

14. Which of the following statements is not true about crossway strip fabric on
    A. Crossway strip is cut at an angle of 45°
    B. Crossway strip is cut at any angle other than 45°.

15. Choose the correct way to make a bound opening decorative.
    A. Using a crossway strip that contrast with the garment
    B. Use zigzag stitch
    C. Use a contrasting thread for the bound opening

16. Select areas where press studs are not applied.
    A. Shoulder and neck openings on children wear
    B. Trousers waist band
    C. Wrist and similar openings and as a finish for main button opening

17. Which one of the following statements is true about one piece cuff and two piece cuffs?
    A. A one piece cuff is cut out from fabric as one piece to be folded into two only half of it will be interfaced while the two pieces is cut as two equal pieces while one becomes the cuff facing and the other is interfaced.
    B. A one piece cuff is cut out on a single layer of fabric while is cut on two layers of fabric

18. Choose the garment which pin stitches are not applied from the following list.
    A. Lingerie
    B. Gents trouser
    C. Children's wear
    D. Fine blouses

19. Faggoting can be used in the following ways except one. Which one?
    A. To attach lace or frills to a hem

B. Attach a collar
C. To make a decorative seam
D. To form an open work or decorative insertion

20. Choose a wrong use of frills
    A. Used for decoration
    B. To lengthen a garment
    C. Reduce opaqueness

21. Which of the following activities is the sequential order for assembling a straight skirt?
    A.
      a) Stitching the darts
      b) Join the back skirts
      c) Stitch the side seams
      d) Attach a waistband
      e) Hem the skirt
      f) Cut and neaten buttonhole
      g) Sew the button
      h) Final pressing
    B.
      a. Stitching the darts
      b. Join the back skirts
      c. Attach a waistband
      d. Hem the skirt
      e. Stitch the side seams
      f. Cut and neaten buttonhole
      g. Sew the button
      h. Stitching the darts
    C.
      a. Hem the skirt
      b. Final pressing
      c. Stitch the side seams
      d. Join the back skirts
      e. Attach a waistband

      f. Cut and neaten buttonhole
      g. Sew the button
      h. Final pressing
D. None of the above

References

Alison Smith. 2009. *The Sewing Book*. UK: Daring Kindersley. <www.dk.com>

M. B. Juvekar. 1992. *Corporal System of Cutting*. Ninth Edition. Bombay: Boox.

Sara May Allington. 1914. *Practical Sewing and Dressmaking*. Third Edition. Boston, USA: The Page Company Publishers.

Valerie I. Cock. 1999. *Dressmaking Simplified*. Third Edition. Malden, MA, USA: Blackwell Science, Inc.

Answers to Revision Questions

Chapter 1

1) C
2) A
3) B
4) C
5) C
6) E
7) B
8) B
9) B
10) D
11) D
12) C
13) D
14) C
15) D
16) B
17) D
18) B

Chapter 2

1. C
2. A
3. C
4. A
5. B
6. D
7. C
8. C
9. A
10. C
11. D

12. B
13. B

Chapter 3

1.
   a. Weight of the fabric being used
   b. Position and purpose of seam
   c. Type and purpose of garment
   d. The style line of a seam
2.
   a. French seam
   b. Run and fell seam
3.
   a. Edge stitching
   b. Overlocking
   c. Binding
   d. Use of pinking shears

Chapter 4

1. Arrangement of the extra material in a garment
2.
   a. Pin tucks
   b. Shell tucks
3. B
4. D
5. C
6. B

Chapter 5

1. Narrow strips of fabric cut at an angle of 45° to the warp and weft threads
2.
    i   Seam or hem edges
    ii   Curved raw edges, armholes and necklines
    iii When attaching collars
    iv To give a decorative finish.
3. Specifically prepared fabric which is set into the garment to give added strength and support areas of strain.
4.
    i. Shaped facing
    ii. Straight facing
    iii. Extended facing
5.
    a. Tailor's canvas
    b. Muslin, scrim and tarlatan
    c. Boded fabrics
6.
    a. Under collar
    b. Cuff edge
    c. On the bodice

Chapter 6

1.
    a. Faced opening
    b. Bound opening
    c. Continuous strap opening
2. Using a crossway strip that contrast with the garment

3.
   a. Back neck openings
   b. Long sleeve wrist openings
4.
a. Skirts
b. Dresses
c. Shirts
d. Jackets
e. Pants

Chapter 7

1.

i Zip fasteners
ii Button and buttonholes
iii Press studs
iv Hooks and eyes

2.
i. Fasteners must be sewn on double fabric
ii. Fasteners must be as inconspicuous as possible unless used as decoration
iii. Attention must be given to details, stitched accurately, and strength of the fastener

3.
   i. Visual method
   ii. Semi-concealed method
   iii. Concealed method
   iv. Invisible method

4.
   a. Worked buttonholes
   b. Bound buttonholes

5.
   a. Jackets

    b.   Cardigans
    c.   Coats

6.
a. Position of the opening
b. Fabric weight
c. Size of the button

Chapter 8

1.
i   Patch pockets
ii Welt pockets

2.
- Complete the top edge, as required.
- Fold the pocket turnings to the wrong side, and press and trim corners to 0.6 cm, snip turnings on curved edges to reduce bulk.
- Place the pocket in position on the garment matching the balance, marks start machine stitching by working the support of the corner first then proceed around the edge of pocket to finish with the support of the second corner.

3.
 i. Self-neatening collar
 ii. Attaching a collar with the use of shaped facing
 iii. Attaching a collar with the use of a crossway strip

4.
A pocket inserted into a garment with the opening strengthened by an added welt

5.
To reduce the bulk

6.
Part of a garment that encircles the neck and frames the face.

7.
    a.   Flat collars
    b.   Standing collars

    c.   Collars cut in one with the garment
    d.   Collars with revers

8.
- a. Polo collar
- b. Shirt collar
- c. Mandarin collar

Chapter 9

1.
   - i   Set-in sleeves
   - ii Puff sleeve
   - iii Petal sleeve
   - iv Flared sleeve
   - v Cape sleeve
   - vi Kimono sleeve
   - vii Raglan sleeve

2.
   - a. Set in sleeves
   - b. Sleeves at are combined with part or entire bodice

3.
   - a. Raglan sleeve
   - b. Kimono sleeve

4. To ensure that fullness is evenly distributed

Chapter 10

1.
   - i Use of applied waistband of self-fabric
   - ii Petersham set inside of skirt

2.
   - i Skirt waistband should be cut along a straight grain to prevent stretching length of the waistband.

ii   Cut waistband to be waist measurements plus 7.5 cm for overlap and turnings twice finished width plus 1.25 cm turnings

3.
- i   Pointed extension of overlap
- ii   Square end with overlap

4.
It is stronger and less easily stretched

5.
To allow for length adjustment and the children grows

6.
- i   Edge stitched hem
- ii   Bound hem
- iii   Slip hemming

Chapter 11

1.
- b.   The garment's fabric
- c.   The style
- d.   Purpose of the garment

2.
a) Shell edging
b) Pin stitch
c) Face of scallops
d) Faggoting
e) Rouleau
f) Frills
g) Lace
h) Braiding

3. Method of linking together two pieces of fabric with a decorative stitch to give an open finish

4.

i. Required length plus half as much
ii. Twice the required length
iii. Three times the required length

Answers to additional questions

1. B
2. B
3. B
4. A
5. C
6. A
7. C
8. D
9. C
10. B
11. A
12. A
13. A
14. B
15. A
16. B
17. B
18. B
19. B
20. C
21. A

About the Author

Morang'a Erick Moseti graduated from Kenya Technical Trainers College (KTTC) in Nairobi, Kenya, and has been a trainer for dressmaking in the past 6 years at the Kolping Vocational Training Centre, Kilimambogo, Kiambu County, Kenya.

Kipchumba Foundation

P.O. Box 25380 – 00100 Nairobi, Kenya

www.kipchumbafound.org

www.ingramcontent.com/pod-product-compliance
Lightning Source LLC
Chambersburg PA
CBHW021811170526
45157CB00007B/2537